D1223128

Mathematics for Everyday Life 12

Authors
Enzo Carli, Sandra Emms Jones,
Alexis Galvao, Peter Joong,
Arnie Niederhoffer, Loraine Wilson

Director of Publishing
David Steele

Publisher
Jan Elliott

Program Manager
Janice Nixon

Developmental Editors
Janice Nixon, Mary Reeve

Copy/Production Editor
Kate Revington

Production Coordinator
Sharon Latta Paterson

Cover and Text Design
Dave Murphy/ArtPlus Ltd.

Page Layout
Karen Wolfe/ArtPlus Ltd.

Technical Art
ArtPlus Ltd.

Photographer
Trent Photographics

Permissions/Photo Research
Lisa Brant

Printer
Transcontinental Printing Inc.

Printed and bound in Canada
1 2 3 4 05 04 03 02

For more information contact Nelson, 1120 Birchmount Road, Toronto, Ontario, M1K 5G4. Or you can visit our Internet site at http://www.nelson.com

National Library of Canada Cataloguing in Publication

Mathematics for everyday life 12 / Enzo Carli ... [et al.]

Includes index.
ISBN 0-7725-2928-0

1. Finance, Personal—Mathematics. I. Carli, E. G.

QA39.2.M2933 2002
332.024'001'513
C2002-905468-0

We acknowledge for their financial support of our publishing program the Canada Council, the Ontario Arts Council, and the Government of Canada through the Book Publishing Industry Development Program (BPIDP).

Acknowledgements

The authors and editors of *Mathematics for Everyday Life 12* wish to thank the reviewers listed below for their valuable input and assistance in ensuring that this text meets the needs of teachers and students in Ontario.

Tom Chapman
Hastings and Prince Edward District School Board

Chris Dearling
Burlington

Linda Palmason
Kawartha Pine Ridge District School Board

Peter Saarimaki
Scarborough

Shirley Scott
District School Board of Niagara

Susan K. Smith
Peel District School Board

Contents

About Your Textbook vi

Chapter 1 *Data Graphs*

1.1 Interpreting Graphs 2
1.2 Constructing Graphs 10
1.3 Career Focus: Factory Worker 14
1.4 Constructing and Interpreting Graphs 16
1.5 Misleading Graphs 20
1.6 Putting It All Together: Selling Shoes 24
1.7 Chapter Review 26

Chapter 2 *Collecting and Organizing Data*

2.1 Sampling Techniques 30
2.2 Statistics in the Media 33
2.3 Organizing and Interpreting Data 35
2.4 Career Focus: Telemarketer 39
2.5 Putting It All Together: Conducting a Survey 41
2.6 Chapter Review 42

Chapter 3 *Probability*

3.1 Making Predictions 46
3.2 Making Decisions 48
3.3 Career Focus: Medical Office Receptionist 50
3.4 Comparing Probabilities 52
3.5 Probability Experiments 54
3.6 Simulations 56
3.7 Putting It All Together: Spring Birthdays 59
3.8 Chapter Review 60

Chapter 4 *Renting an Apartment*

4.1 Availability of Apartments 64
4.2 Renting an Apartment 67
4.3 Rights and Responsibilities of Landlords and Tenants 71
4.4 Career Focus: Building Superintendent 75
4.5 Monthly Apartment Costs 76
4.6 Putting It All Together: Renting an Apartment 79
4.7 Chapter Review 80

Chapter 5 *Buying a Home*

5.1 Looking for a Home 84
5.2 Buying a Home 86
5.3 The Costs of Maintaining a Home 91
5.4 Career Focus: Real Estate Agent 94
5.5 Putting It All Together: Buying a Home 96
5.6 Chapter Review 97

Chapter 6 *Household Budgets*

6.1 Affordable Housing 100
6.2 Components of a Household Budget 105
6.3 Monthly Budget 108
6.4 Changing One Item in a Budget 115
6.5 Career Focus: Furniture Refinisher 118
6.6 Putting It All Together: Household Budgets 119
6.7 Chapter Review 120

Chapter 7 *Measuring and Estimating*

7.1 The Metric System 124
7.2 Measuring Lengths 127
7.3 Estimating Distances 130
7.4 Estimating Capacities 132
7.5 Estimating Large Numbers 134
7.6 Career Focus: Decorating Store Clerk 136
7.7 Putting It All Together: Estimating 137
7.8 Chapter Review 138

Chapter 8 *Measurement and 2-D Design*

8.1 The Pythagorean Theorem 142
8.2 Calculating Perimeter and Area 144
8.3 Estimating Perimeter and Area 148
8.4 Career Focus: Flooring Installer 150
8.5 Enlargements 151
8.6 Scale Drawings 153
8.7 Putting It All Together: Designing a Playground 155
8.8 Chapter Review 156

Chapter 9 *Measurement and 3-D Design*

9.1 Rectangular Prisms 160
9.2 Cylinders 164
9.3 3-D Drawings 168
9.4 Scale Models 169
9.5 Career Focus: Hobby Store Clerk 171
9.6 Putting It All Together: Finishing a Basement 172
9.7 Chapter Review 173

Chapter 10 *Transformations and Design*

10.1 Geometric Aspects of Design 176
10.2 Investigating Design Using Technology 181
10.3 Designing a Logo 185
10.4 Career Focus: Sign Painter 186
10.5 Tiling a Plane 187
10.6 Designs Involving Tiling Patterns 189
10.7 Putting It All Together: Designing Geometrically 190
10.8 Chapter Review 191

Glossary 192

Answers 196

Index 219

Credits 220

About Your Textbook

When you turn the pages of your new textbook, you will notice that most sections begin with **Explore**. Here, you usually work in a small group or with a partner. You are given an opportunity to connect your prior math knowledge and your personal experience to concepts in the section. What you have to do varies in scope from briefly reflecting and discussing to applying the steps of an inquiry/problem solving process.

An **inquiry/problem solving process** involves
- formulating questions
- selecting strategies, resources, technology, and tools
- representing in mathematical form
- interpreting information and forming conclusions
- reflecting on the reasonableness of results

Most sections continue with **Develop**.

Here, you sometimes work in a small group or with a partner. The concepts of the section are developed by presenting you with
- directed questions to answer and/or
- examples and fully worked solutions to follow

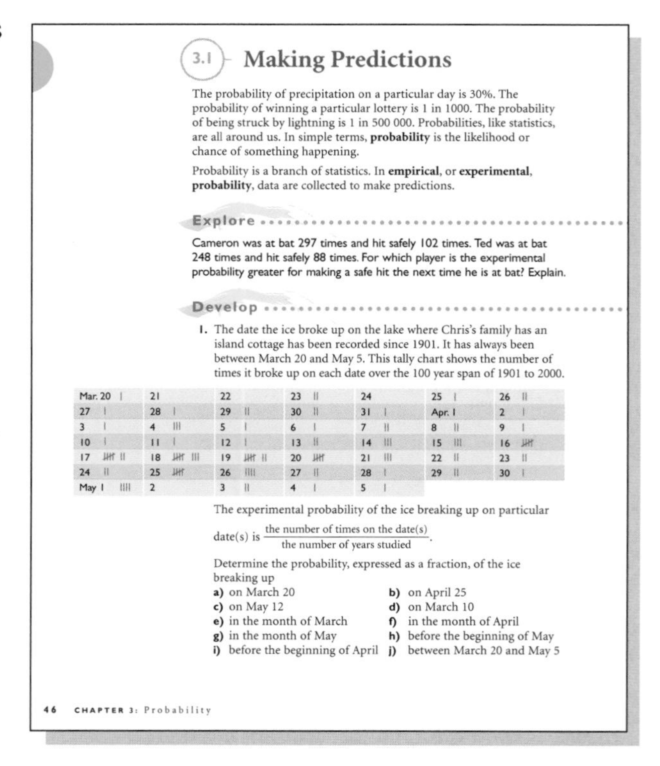

3.1 Making Predictions

The probability of precipitation on a particular day is 30%. The probability of winning a particular lottery is 1 in 1000. The probability of being struck by lightning is 1 in 500 000. Probabilities, like statistics, are all around us. In simple terms, **probability** is the likelihood or chance of something happening.

Probability is a branch of statistics. In **empirical**, or **experimental**, **probability**, data are collected to make predictions.

Explore

Cameron was at bat 297 times and hit safely 102 times. Ted was at bat 248 times and hit safely 88 times. For which player is the experimental probability greater for making a safe hit the next time he is at bat? Explain.

Develop

1. The date the ice broke up on the lake where Chris's family has an island cottage has been recorded since 1901. It has always been between March 20 and May 5. This tally chart shows the number of times it broke up on each date over the 100 year span of 1901 to 2000.

<table>
<tr><td>Mar. 20 |</td><td>21</td><td>22</td><td>23 ||</td><td>24</td><td>25 |</td><td>26 ||</td></tr>
<tr><td>27 |</td><td>28 |</td><td>29 ||</td><td>30 ||</td><td>31 |</td><td>Apr. 1</td><td>2 |</td></tr>
<tr><td>3 |</td><td>4 |||</td><td>5 |</td><td>6 |</td><td>7 ||</td><td>8 ||</td><td>9 |</td></tr>
<tr><td>10 |</td><td>11 |</td><td>12 |</td><td>13 ||</td><td>14 |||</td><td>15 |||</td><td>16 卌</td></tr>
<tr><td>17 卌 ||</td><td>18 卌 |||</td><td>19 卌 ||</td><td>20 卌</td><td>21 |||</td><td>22 ||</td><td>23 ||</td></tr>
<tr><td>24 ||</td><td>25 卌</td><td>26 ||||</td><td>27 ||</td><td>28 |</td><td>29 ||</td><td>30 |</td></tr>
<tr><td>May 1 ||||</td><td>2</td><td>3 ||</td><td>4 |</td><td>5 |</td><td></td><td></td></tr>
</table>

The experimental probability of the ice breaking up on particular date(s) is $\frac{\text{the number of times on the date(s)}}{\text{the number of years studied}}$.

Determine the probability, expressed as a fraction, of the ice breaking up

a) on March 20
b) on April 25
c) on May 12
d) on March 10
e) in the month of March
f) in the month of April
g) in the month of May
h) before the beginning of May
i) before the beginning of April
j) between March 20 and May 5

46 CHAPTER 3: Probability

Use the probabilities you determined in question 1 to answer questions 2 and 3. Explain your answers.

2. **a)** Would you believe that the ice broke up on June 1 in 2001?
b) Would you believe that the ice broke up on March 15 in 2002?
c) Which is the most probable date from question 1 parts a), b), and c) that the ice broke up on in 2003?
d) How might Chris's family use this information?

3. **a)** What do you notice about the answers to question 1 parts e) and i)?
b) What do you notice about the answers to question 1 parts c) and d)? What is something in your everyday life that has the same probability?
c) What do you notice about the answer to question 1 part j)? What is something in your everyday life that has the same probability?
d) In which month is the probability the greatest that the ice will break up?

Practise

4. **Skills Check** Use your knowledge of decimals to list the batting averages from the most probable to make a safe hit next time at bat to the least probable.
0.284 0.373 0.432 0.401 0.337 0.291 0.410

5. A fitness study asked 300 people between the ages of 15 and 74 if they exercised at least three times a week. The yes responses were tallied by age groupings.

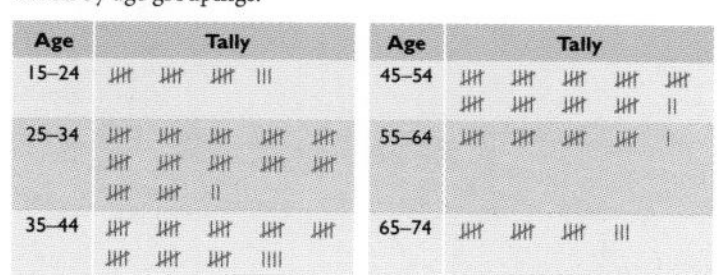

Age	Tally	Age	Tally
15–24	卌 卌 卌 III	45–54	卌 卌 卌 卌 卌 卌 卌 卌 卌 II
25–34	卌 卌 卌 卌 卌 卌 卌 卌 卌 卌 卌 卌 II	55–64	卌 卌 卌 卌 I
35–44	卌 卌 卌 卌 卌 卌 卌 卌 IIII	65–74	卌 卌 卌 III

Determine the experimental probability, expressed as a fraction, that the response would be yes if the same question was asked of one more person
a) aged 15 to 24 **b)** aged 35 to 44
c) over 24 **d)** under 45

6. How would the information in question 5 be used by each of the following?
a) a government agency advertising the importance of fitness
b) a fitness club advertising their facilities

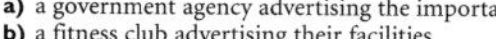

SECTION 3.1: Making Predictions 47

Most sections then have **Practise**. Here, you frequently work alone to answer questions about the skills and concepts presented in the section.

In each chapter you will find one of each of the following special sections.

Chapter Opener

This introduction outlines what you are expected to accomplish in the chapter.

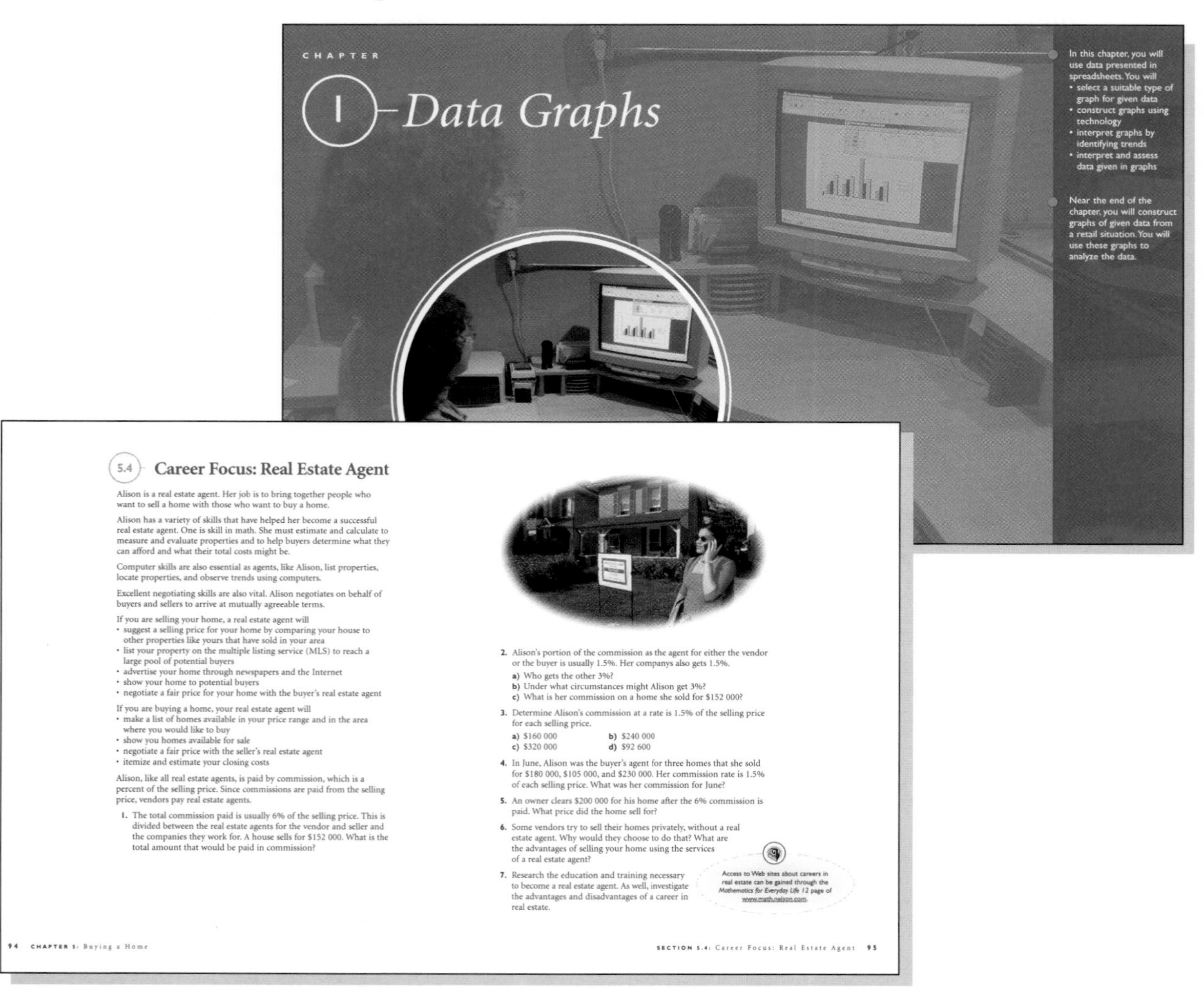

CHAPTER

1 Data Graphs

In this chapter, you will use data presented in spreadsheets. You will
- select a suitable type of graph for given data
- construct graphs using technology
- interpret graphs by identifying trends
- interpret and assess data given in graphs

Near the end of the chapter, you will construct graphs of given data from a retail situation. You will use these graphs to analyze the data.

5.4 Career Focus: Real Estate Agent

Alison is a real estate agent. Her job is to bring together people who want to sell a home with those who want to buy a home.

Alison has a variety of skills that have helped her become a successful real estate agent. One is skill in math. She must estimate and calculate to measure and evaluate properties and to help buyers determine what they can afford and what their total costs might be.

Computer skills are also essential as agents, like Alison, list properties, locate properties, and observe trends using computers.

Excellent negotiating skills are also vital. Alison negotiates on behalf of buyers and sellers to arrive at mutually agreeable terms.

If you are selling your home, a real estate agent will
- suggest a selling price for your home by comparing your house to other properties like yours that have sold in your area
- list your property on the multiple listing service (MLS) to reach a large pool of potential buyers
- advertise your home through newspapers and the Internet
- show your home to potential buyers
- negotiate a fair price for your home with the buyer's real estate agent

If you are buying a home, your real estate agent will
- make a list of homes available in your price range and in the area where you would like to buy
- show you homes available for sale
- negotiate a fair price with the seller's real estate agent
- itemize and estimate your closing costs

Alison, like all real estate agents, is paid by commission, which is a percent of the selling price. Since commissions are paid from the selling price, vendors pay real estate agents.

1. The total commission paid is usually 6% of the selling price. This is divided between the real estate agents for the vendor and seller and the companies they work for. A house sells for $152 000. What is the total amount that would be paid in commission?

2. Alison's portion of the commission as the agent for either the vendor or the buyer is usually 1.5%. Her companys also gets 1.5%.
   a) Who gets the other 3%?
   b) Under what circumstances might Alison get 3%?
   c) What is her commission on a home she sold for $152 000?

3. Determine Alison's commission at a rate is 1.5% of the selling price for each selling price.
   a) $160 000 b) $240 000
   c) $320 000 d) $92 600

4. In June, Alison was the buyer's agent for three homes that she sold for $180 000, $105 000, and $230 000. Her commission rate is 1.5% of each selling price. What was her commission for June?

5. An owner clears $200 000 for his home after the 6% commission is paid. What price did the home sell for?

6. Some vendors try to sell their homes privately, without a real estate agent. Why would they choose to do that? What are the advantages of selling your home using the services of a real estate agent?

7. Research the education and training necessary to become a real estate agent. As well, investigate the advantages and disadvantages of a career in real estate.

Access to Web sites about careers in real estate can be gained through the *Mathematics for Everyday Life 12* page of www.math.nelson.com.

94 CHAPTER 5: Buying a Home

SECTION 5.4: Career Focus: Real Estate Agent 95

Career Focus

In this section you read and answer questions about a job which can be entered right after high school. The questions that you answer also offer practice in some of the Essential Skills—

- text reading
- document use
- writing
- numeracy (math)
- working with others
- continuous learning
- computer skills
- thinking skills

Developing these skills will enable you to participate fully in the workplace and the community.

Putting It All Together

In this section you are given an opportunity to demonstrate skills in the interconnected concepts of the chapter. Sometimes, you are presented with specific questions to answer. Other times, you are presented with a problem that calls upon you to apply the steps of an inquiry/problem solving process.

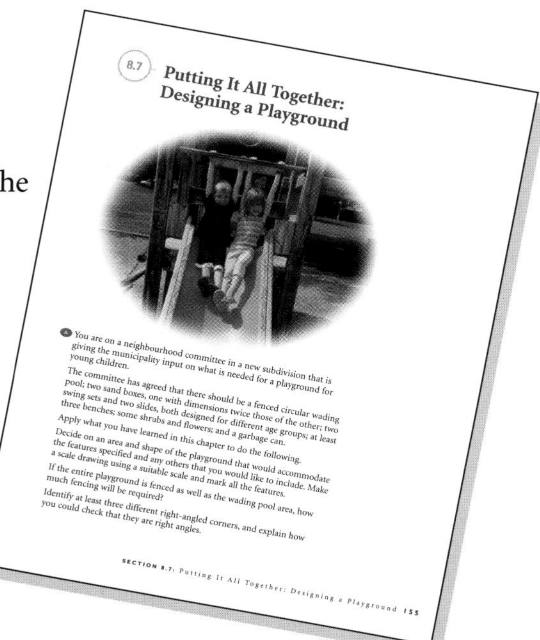

8.7 Putting It All Together: Designing a Playground

You are on a neighbourhood committee in a new subdivision that is giving the municipality input on what is needed for a playground for young children.

The committee has agreed that there should be a fenced circular wading pool; two sand boxes, one with dimensions twice those of the other; two swing sets and two slides, both designed for different age groups; at least three benches; some shrubs and flowers; and a garbage can.

Apply what you have learned in this chapter to do the following.

Decide on an area and shape of the playground that would accommodate the features specified and any others that you would like to include. Make a scale drawing using a suitable scale and mark all the features.

If the entire playground is fenced as well as the wading pool area, how much fencing will be required?

Identify at least three different right-angled corners, and explain how you could check that they are right angles.

SECTION 8.7: Putting It All Together: Designing a Playground 155

Chapter Review

Here, you find questions similar to the Practise questions in earlier sections. Answering them helps you prepare for a chapter test.

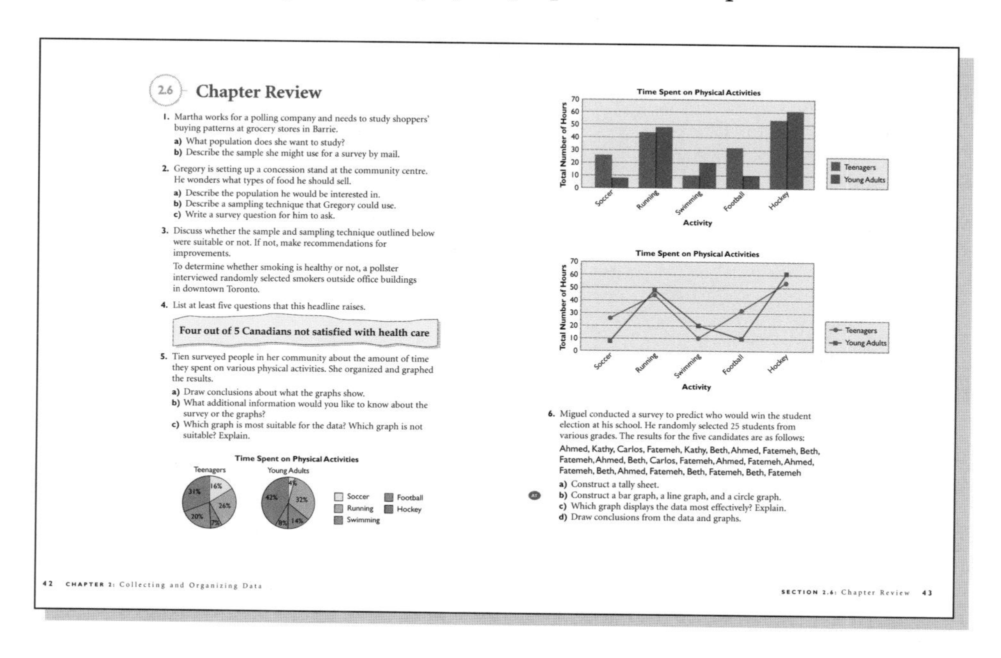

2.6 Chapter Review

1. Martha works for a polling company and needs to study shoppers' buying patterns at grocery stores in Barrie.
   a) What population does she want to study?
   b) Describe the sample she might use for a survey by mail.
2. Gregory is setting up a concession stand at the community centre. He wonders what types of food he should sell.
   a) Describe the population he would be interested in.
   b) Describe a sampling technique that Gregory could use.
   c) Write a survey question for him to ask.
3. Discuss whether the sample and sampling technique outlined below were suitable or not. If not, make recommendations for improvements.
   To determine whether smoking is healthy or not, a pollster interviewed randomly selected smokers outside office buildings in downtown Toronto.
4. List at least five questions that this headline raises.
   Four out of 5 Canadians not satisfied with health care
5. Tien surveyed people in her community about the amount of time they spent on various physical activities. She organized and graphed the results.
   a) Draw conclusions about what the graphs show.
   b) What additional information would you like to know about the survey or the graphs?
   c) Which graph is most suitable for the data? Which graph is not suitable? Explain.

6. Miguel conducted a survey to predict who would win the student election at his school. He randomly selected 25 students from various grades. The results for the five candidates are as follows:
   Ahmed, Kathy, Carlos, Fatemeh, Kathy, Beth, Ahmed, Fatemeh, Beth, Fatemeh, Ahmed, Beth, Carlos, Fatemeh, Ahmed, Fatemeh, Ahmed, Fatemeh, Beth, Ahmed, Fatemeh, Beth, Fatemeh, Beth, Fatemeh
   a) Construct a tally sheet.
   b) Construct a bar graph, a line graph, and a circle graph.
   c) Which graph displays the data most effectively? Explain.
   d) Draw conclusions from the data and graphs.

42 CHAPTER 2: Collecting and Organizing Data

SECTION 2.6: Chapter Review 43

2. Rather than drawing all those cards, you can simulate the situation using a random number generator. Because you need four equally likely outcomes for the four possible answers, generate random integers from 1 to 4. Because there are ten questions, generate ten random numbers at a time.
   a) Prepare the random integer generator.
   b) Decide which number will represent getting a correct answer.
   c) Create a tally chart with space for five trials as in question 1.
   d) Record the first seven numbers that are visible as correct or wrong answers. Then, scroll through the numbers until you reach the closing bracket. Record the last three numbers that you could not initially see as correct or wrong answers.
   e) Press enter and repeat part d) for a second trial. Continue until you have five trials.
   f) What experimental probability of passing by guessing did you determine this time?
   g) Combine your results from part e) with those of your classmates. Is this experimental probability closer to the theoretical probability from question 1 i) than your probability in part f)?

Practise

3. Use a simulation to determine the probability that in a family with four children, all four are boys.
   a) How many equally likely outcomes are there for the gender of a child?
   b) Select a physical model to use.
   c) Decide which outcome will represent a boy.
   d) Create a tally chart with space for ten trials.
   e) Perform ten trials.
   f) What is the experimental probability that in a family with four children, all four are boys?
   g) Combine your results from part e) with those of your classmates.
   h) Would you believe that the theoretical probability of all four children in a family being boys is 0.0625? Explain.

SECTION 3.6: Simulations 57

2. Kyle faces a rent increase of 3.9% in two months. His current monthly budget is shown. Revise his budget to deal with the upcoming rent increase.

	A	B
1	**Monthly Net Income**	**Budget ($)**
2	**Total**	2281
3	**Monthly Expenses**	
4	**Fixed expenses**	
5	savings	290
6	rent	650
7	parking	20
8	car payment	343
9	car insurance	120
10	cable	48
11	**Variable expenses**	
12	transportation—gas, oil	120
13	food	360
14	telephone	40
15	clothing	100
16	entertainment	120
17	personal items	50
18	household items	20
19	**Total monthly expenses**	2281

3. During a slow time at Kyle's workplace, all employees had their hours cut to 32 from 40. Kyle's monthly net income was reduced to $1824.80 from $2281. Refer to Kyle's budget in question 2. Suggest several items on which Kyle could cut back to ensure that his reduced income covers his expenses, including the increased rent. Kyle does not want to move. Create a revised budget for Kyle.

Develop

1. Use the Internet to access The Ontario Rental Housing Tribunal Web site or use brochures from The Ontario Rental Housing Tribunal to answer the Renter's Quiz again. Are you surprised by any of the answers?

Access to The Ontario Rental Housing Tribunal Web site can be gained through the *Mathematics for Everyday Life 12* page of www.math.nelson.com.

Practise

2. When a landlord rents to a person, the landlord gives the person a **tenancy**, or a legal right to occupy the rental unit. Every tenant and landlord have a **tenancy agreement**, which may or may not be in writing. A written tenancy agreement is often called a **lease**. When a tenancy agreement is for a specified period of time with start and end dates set out, the tenancy has a **fixed term**.
   a) What is meant by leasing for a year?
   b) What would renting month-to-month mean?
   c) Rent may be due weekly, biweekly, or monthly. Determine the yearly rent for each apartment described below.
      i) $825 per month
      ii) $190 per week
      iii) $400 biweekly
   d) The three apartments in part c) are the same size and have the same features. Why might someone rent the most expensive apartment?
3. **Skills Check** Calculate each percent, rounded to the nearest cent.
   a) 2.5% of $1100 monthly rent
   b) 3.4% of $800 monthly rent
   c) 2.9% of $750 monthly rent
   d) 2.8% of $675 monthly rent
   e) 3.1% of $1100 monthly rent
   f) 3.7% of $820 monthly rent
4. The Ontario government sets maximum rent increases each year.

Year	Percent Increase
2000	2.6
2001	2.9
2002	3.9
2003	2.9

68 CHAPTER 4: Renting an Apartment

Technology in *Mathematics for Everyday Life 12*

Technology plays an important role in supporting your learning.

You will use
- calculators to perform operations with decimals
- spreadsheets to draw graphs to display data, and to present and change budgets
- graphing calculators to generate random numbers
- dynamic geometry software to create two-dimensional scale drawings, to create three-dimensional drawings, and to investigate transformations

Using these tools will enable you to focus on the concepts that you are learning about.

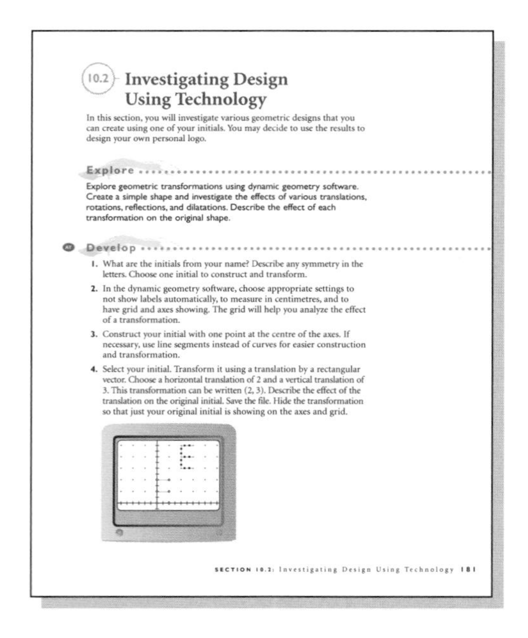

10.2 **Investigating Design Using Technology**

In this section, you will investigate various geometric designs that you can create using one of your initials. You may decide to use the results to design your own personal logo.

Explore

Explore geometric transformations using dynamic geometry software. Create a simple shape and investigate the effects of various translations, rotations, reflections, and dilatations. Describe the effect of each transformation on the original shape.

Develop

1. What are the initials from your name? Describe any symmetry in the letters. Choose one initial to construct and transform.
2. In the dynamic geometry software, choose appropriate settings to not show labels automatically, to measure in centimetres, and to have grid and axes showing. The grid will help you analyze the effect of a transformation.
3. Construct your initial with one point at the centre of the axes. If necessary, use line segments instead of curves for easier construction and transformation.
4. Select your initial. Transform it using a translation by a rectangular vector. Choose a horizontal translation of 2 and a vertical translation of 3. This transformation can be written (2, 3). Describe the effect of the translation on the original initial. Save the file. Hide the transformation so that just your original initial is showing on the axes and grid.

SECTION 10.2: Investigating Design Using Technology 181

You will also use the Internet, as well as print sources, such as newspapers, flyers, pamphlets, and brochures about a variety of topics, to research current information.

Watch for the following.

AT This indicates that an **Alternative to Technology** master is available. Your teacher may provide it if your class is unable to work on computers with spreadsheet software.

This tells you that you can get linked to relevant **Web sites** by going to the *Mathematics for Everyday Life 12* page at www.math.nelson.com.

WS This means that an organizational **Work Sheet** master is available. Your teacher may provide it to help you organize your work for a particular question or group of questions.

A This tells you that an **Assessment** master with a rubric is available. Your teacher may provide it to help you understand how your work might be assessed.

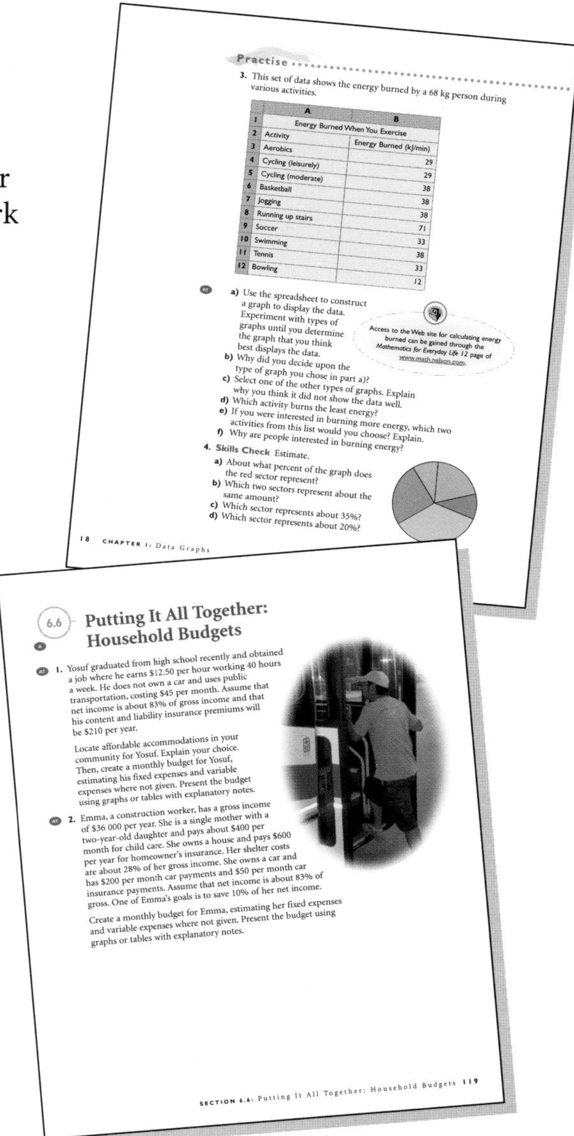

Practise

3. This set of data shows the energy burned by a 68 kg person during various activities.

	A	B
1	Energy Burned When You Exercise	
2	Activity	Energy Burned (kJ/min)
3	Aerobics	29
4	Cycling (leisurely)	29
5	Cycling (moderate)	38
6	Basketball	38
7	Jogging	38
8	Running up stairs	71
9	Soccer	33
10	Swimming	38
11	Tennis	33
12	Bowling	12

a) Use the spreadsheet to construct a graph to display the data. Experiment with types of graphs until you determine the graph that you think best displays the data.
b) Why did you decide upon the type of graph you chose in part a)?
c) Select one of the other types of graphs. Explain why you think it did not show the data well.
d) Which activity burns the least energy?
e) If you were interested in burning more energy, which two activities from this list would you choose? Explain.
f) Why are people interested in burning energy?

Access to the Web site for calculating energy burned can be gained through the *Mathematics for Everyday Life 12* page of www.math.nelson.com.

4. **Skills Check** Estimate.
a) About what percent of the graph does the red sector represent?
b) Which two sectors represent about the same amount?
c) Which sector represents about 35%?
d) Which sector represents about 20%?

18 CHAPTER 1: Data Graphs

6.6 **Putting It All Together: Household Budgets**

1. Yosuf graduated from high school recently and obtained a job where he earns $12.50 per hour working 40 hours a week. He does not own a car and uses public transportation, costing $45 per month. Assume that net income is about 83% of gross income and that his content and liability insurance premiums will be $210 per year.

Locate affordable accommodations in your community for Yosuf. Explain your choice. Then, create a monthly budget for Yosuf, estimating his fixed expenses and variable expenses where not given. Present the budget using graphs or tables with explanatory notes.

2. Emma, a construction worker, has a gross income of $36 000 per year. She is a single mother with a two-year-old daughter and pays about $400 per month for child care. She owns a house and pays $600 per year for homeowner's insurance. Her shelter costs are about 28% of her gross income. She owns a car and has $200 per month car payments and $50 per month car insurance payments. Assume that net income is about 83% of gross. One of Emma's goals is to save 10% of her net income.

Create a monthly budget for Emma, estimating her fixed expenses and variable expenses where not given. Present the budget using graphs or tables with explanatory notes.

SECTION 6.6: Putting It All Together: Household Budgets 119

Other Features

CD with spreadsheet templates
Templates in Microsoft Excel and Corel Quattro Pro are provided for you every time you need to work with spreadsheets.

Glossary
All the new terms in the book are listed alphabetically. Terms are explained when first introduced, but whenever you need to confirm the meaning of a term, you can turn to the Glossary.

Answers
You will find that answers to all questions are given except when they would vary depending on personal choices. Check your answers against these. If your answer differs, work backward to try to understand how the textbook's answer was determined.

CHAPTER

1 Data Graphs

In this chapter, you will use data presented in spreadsheets. You will
- select a suitable type of graph for given data
- construct graphs using technology
- interpret graphs by identifying trends
- interpret and assess data given in graphs

Near the end of the chapter, you will construct graphs of given data from a retail situation. You will use these graphs to analyze the data.

1.1 Interpreting Graphs

When you want to describe a memorable scene, you can use words. However, a picture often shows it better.

You can arrange data or information in a table. However, a graph provides a quick impression and often makes it easier to see trends.

Explore

Below is information on the average length of time that males were expected to live at various times over the last 200 years. **Statistics Canada** gathered the data. Statistics Canada is the government agency that collects, organizes, and analyzes data about many aspects of life in Canada.

Access to the Statistics Canada Web site can be gained through the *Mathematics for Everyday Life 12* page of www.math.nelson.com.

Life Expectancy at Birth for Males Born from 1801 to 1941

1801: 37.76 years	1841: 40.78 years	1881: 47.95 years	1921: 62.85 years
1811: 38.61 years	1851: 41.71 years	1891: 49.35 years	1931: 66.28 years
1821: 39.40 years	1861: 42.70 years	1901: 53.16 years	1941: 70.66 years
1831: 40.13 years	1871: 45.27 years	1911: 57.70 years	

How could you organize the data? What trends do you see in the data? What conclusions can you draw? What types of graphs could you use to display the data? Which type of graph would show the trends most clearly?

Develop

1. James works for a newspaper. He is writing an article about the cost of clearing snow from cities in Canada. The cost is related to the amount of snowfall. James found this information on the Statistics Canada Web site.

City	Average Annual Snowfall (cm)
Vancouver	54.9
Calgary	135.4
Regina	107.4
Toronto	135.0
Ottawa	221.5
Quebec City	337.0
Halifax	261.4

James decides to use a **bar graph** in his article. Bar graphs are used to compare quantities. Equal spaces separate the bars. Those with vertical bars are sometimes called **column graphs**.

James uses a spreadsheet to draw the graph. He puts labels and a title on the graph. Each bar represents the average quantity of snow in one city.

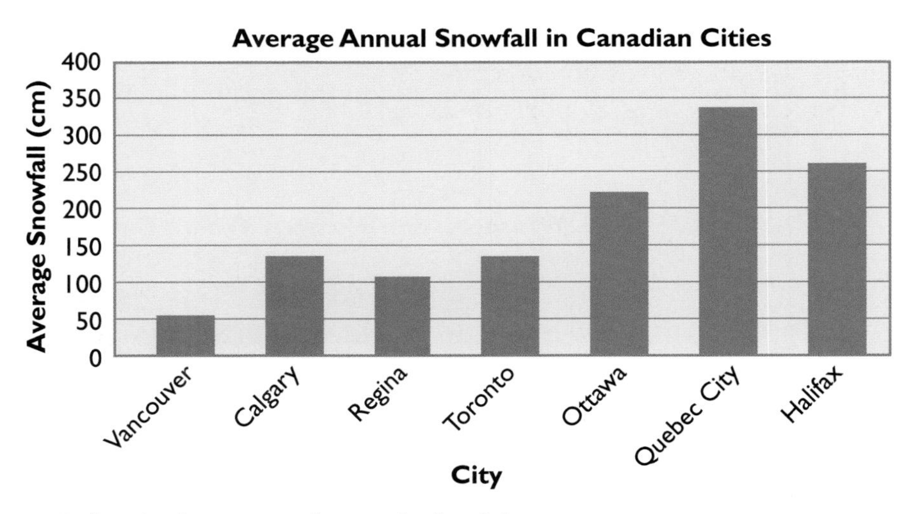

a) What is shown on the vertical axis?
b) Describe the scale on the vertical axis. How was this scale chosen?
c) What is shown on the horizontal axis?
d) Which city has the greatest average snowfall?
e) Which city has the lowest average snowfall?
f) Which two cities are closest in average snowfall?
g) Which way of presenting the data is easier to read—the table above or the bar graph? Explain.

h) James sends the bar graph to the art department at the newspaper. The graphic artist draws the graph so that it catches the attention of readers. What are the advantages of this presentation? the disadvantages?

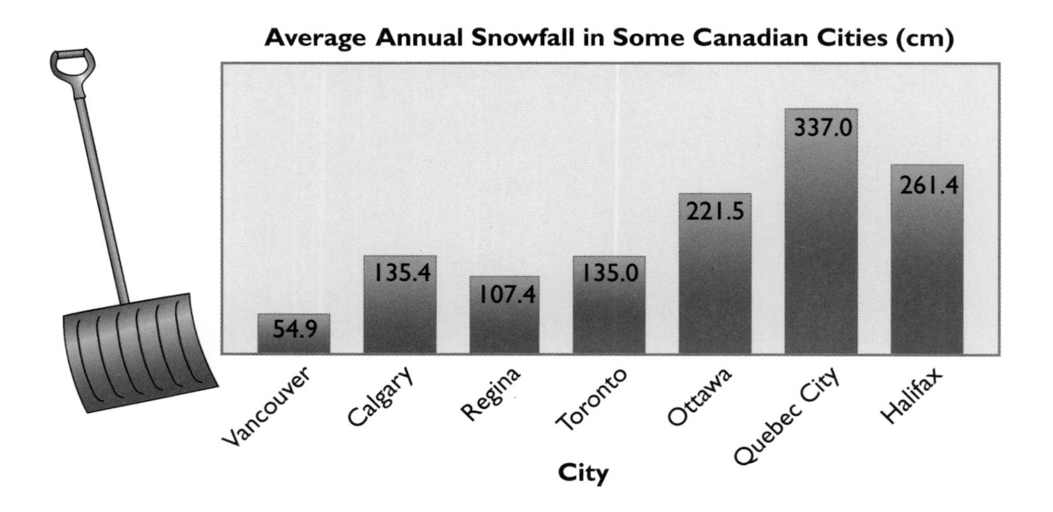

2. A government Web site has information about various jobs. In the section about transportation workers, a bar graph shows in what areas these employees work. The bars are drawn horizontally.

a) What do the numbers at the right of the bars represent?

b) What information might you add to this graph to make it easier to read?

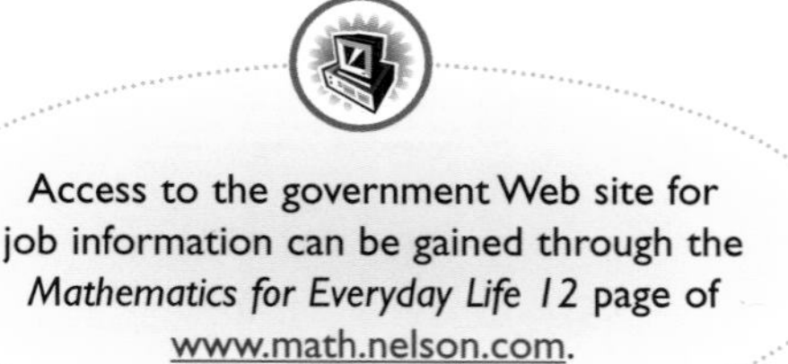

Access to the government Web site for job information can be gained through the *Mathematics for Everyday Life 12* page of www.math.nelson.com.

Transportation Workers: Where They Work

Area	
Business Services	44.3
Air Transportation	18.3
Rail Transportation	14.2
Federal Administration	9.4
Water Transportation	7.0
Public Transit and Other Transportation	2.7
Wholesale Trade	0.6

3. This graph shows the percent of the population in various age groups working in transportation occupations and in all occupations.

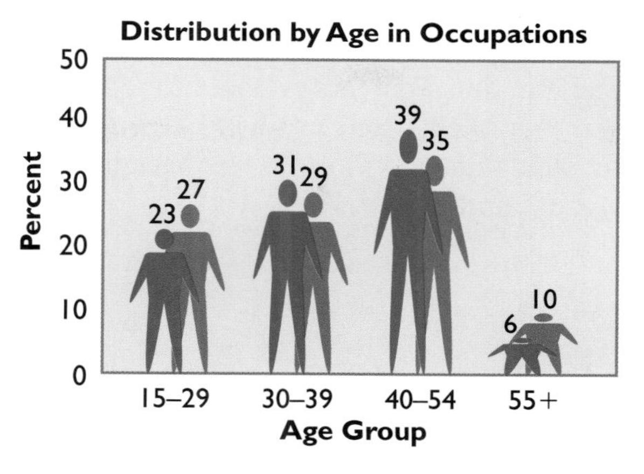

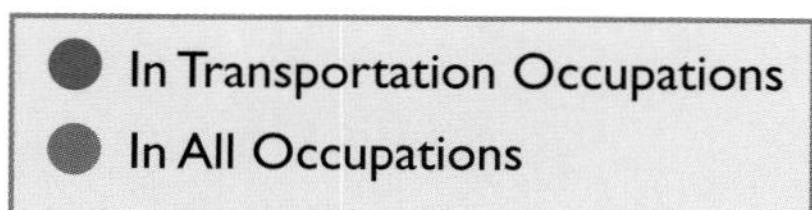

a) What is the total percent of people in the transportation industry?
b) What is the total percent of people in all occupations?
c) Why do you think the totals are not 100%?
d) In which age group are most people who work in the transportation industry?
e) The data could also have been displayed in a **double bar graph**. Double bar graphs are used to compare two similar sets of data. Which graph do you find easier to understand—the above graph or the double bar graph below? Why?

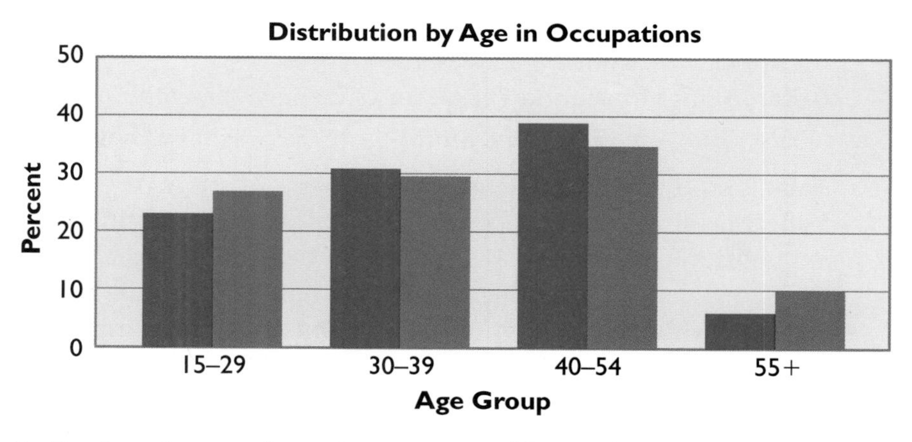

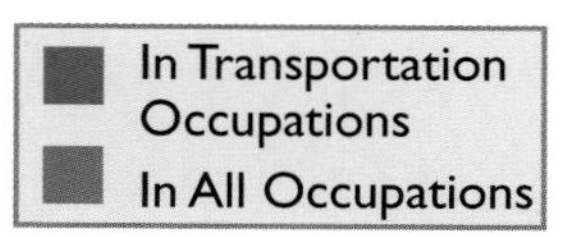

f) Look at the number of years covered by each age group on the graphs. How might this make the graphs misleading?

4. Karen is making a poster for an independent study project for physical education class. She found information in the newspaper on times taken to run the 100 m dash in several years from 1930 to 1999.

Karen uses a spreadsheet to draw a **line graph** to display the information. On a line graph, line segments are used to join data points. Line graphs often show a change over time. Karen adds interest to her graph by changing the colour of the background and adding some clip art of runners.

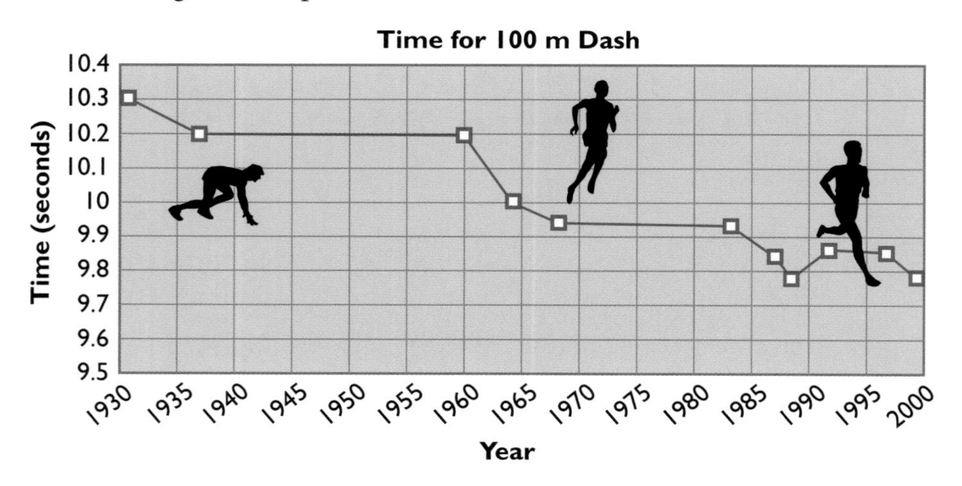

a) What was the fastest time for the 100 m dash in 1964?

b) Between what two years was there the greatest change in the time for the 100 m dash? How do you know?

c) When was there an increase in the time for the 100 m dash? How do you know?

d) Karen might have shown the times using a bar graph. Why do you think she used a line graph?

5. Frank is studying the effect of various forms of stress on people. He thinks that stress due to a combination of poor economic conditions and political unrest may cause people to drive carelessly and to disagree violently with other people. He finds data on the number of people who have died in traffic accidents and the number of murders committed in the same time period.

Frank uses a **double line graph** to display the data. (See next page.) A double line graph is used to compare two similar sets of data that change over time.

a) What trends would Frank look for in the graph?

b) What trends do you notice?

c) If Frank is correct, more people drive carelessly at the same time as more people disagree violently. Do you think the graph shows this to be true? Explain.

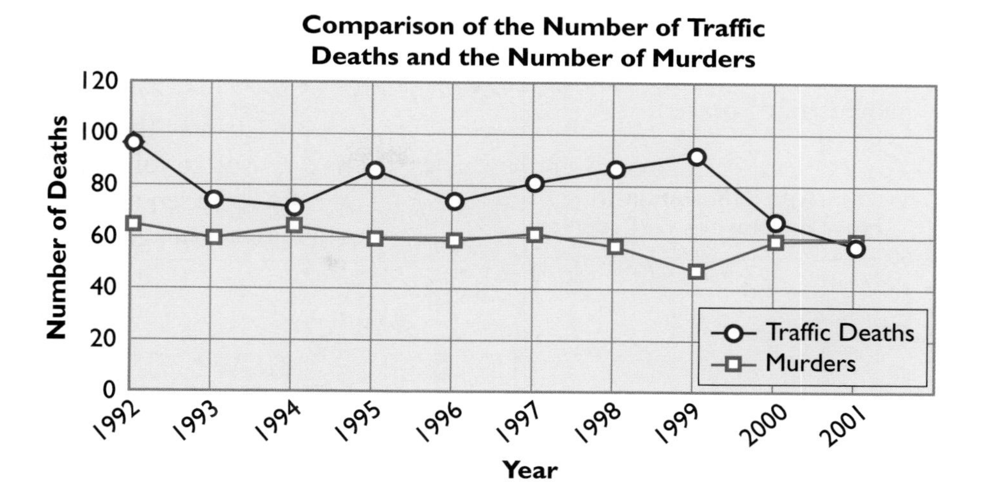

6. A city presented its budget as a **pie graph** (also called a **circle graph**). A pie graph has sectors that represent portions of a whole. In this pie graph for property taxes, each sector represents the property tax paid for one area as a percent of the total collected from property owners.

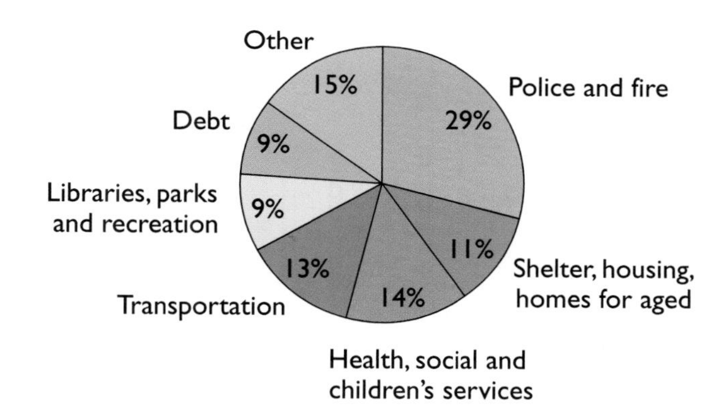

a) Which area receives the greatest percent of taxes?
b) Which two areas receive the same percent of taxes?
c) Which two areas together receive the same as the area in part a)?
d) What is the total of the percents shown on the graph? Why would this be?
e) A homeowner pays a property tax of $3000. How many of these tax dollars are spent on the police and fire departments?

7. Later in this course you are going to create personal and household budgets. To prepare for this, start tracking your spending. Set up a sheet with headings:

Date	Item	Amount

Practise

8. Identify each type of graph. Explain why each graph is suitable for the data displayed.

a)

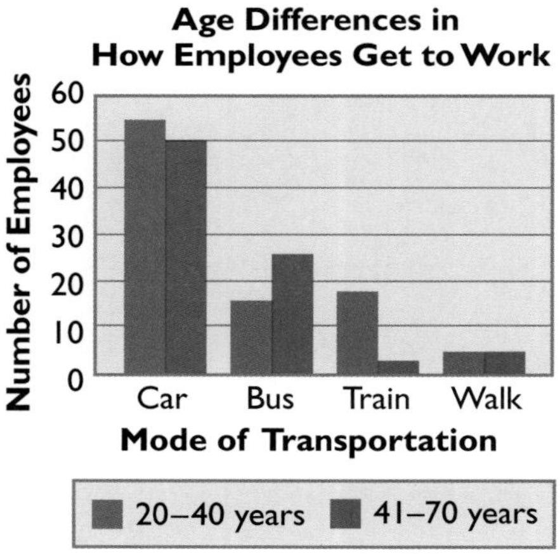

Age Differences in
How Employees Get to Work
Number of Employees
0
10
20
30
40
50
60
Car
Bus
Train
Walk
Mode of Transportation
20–40 years
41–70 years

b)

Temperature in July
Temperature (°C)
0
10
20
30
40
5
10
15
20
Date

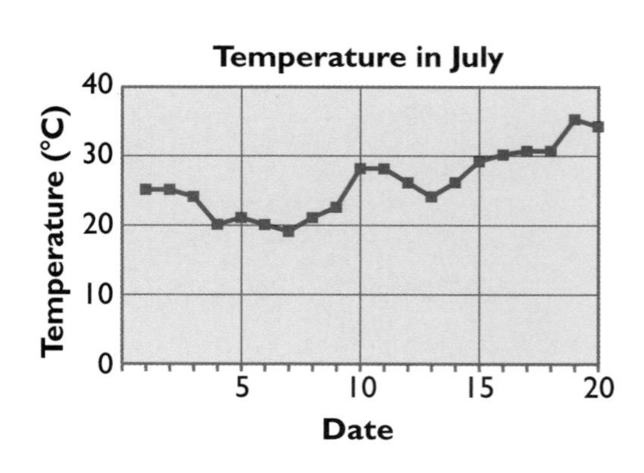

c)

Time Spent on Homework
French
22%
Phys. ed.
11%
Math
33%
History
17%
Novel study
17%

d)

Owen Sound and Yellowknife
Temperature in July
Temperature (°C)
0
10
20
30
40
5
10
15
20
Date
Owen Sound
Yellowknife

e)

How Employees Get to Work
Number of Employees
0
20
40
60
80
100
120
Car
Bus
Train
Walk
Mode of Transportation

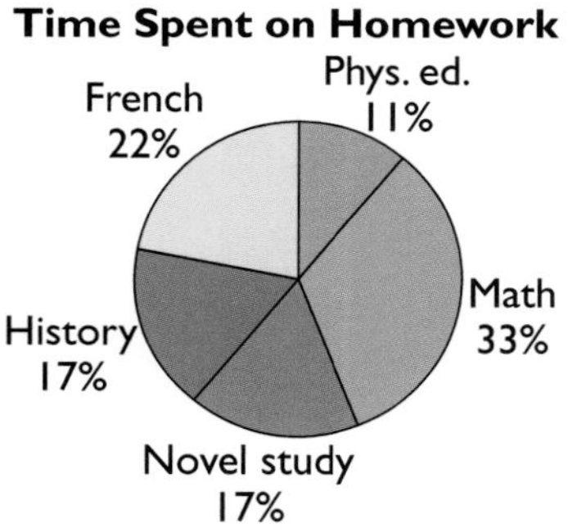

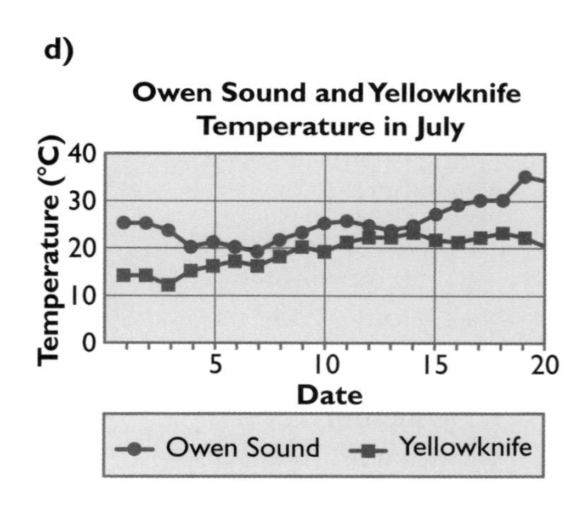

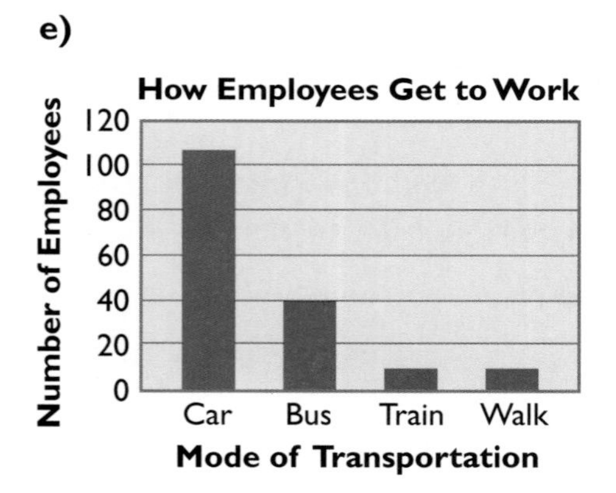

9. The city art department could display the distribution of a certain amount of property taxes in a bar graph, instead of using the pie graph in question 6.

a) How are the pie and bar graphs alike? How are they different?

b) Which graph do you think is more suitable for displaying the data? Explain.

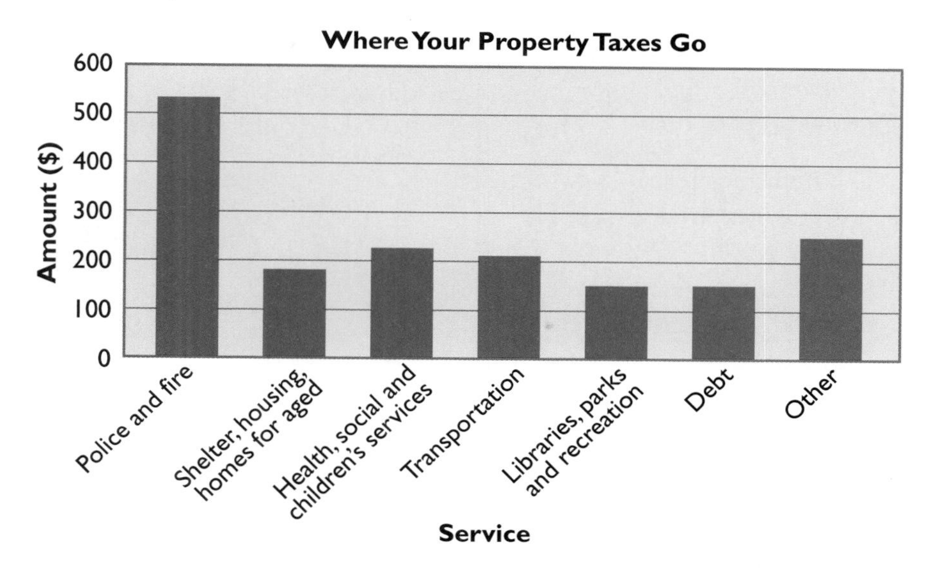

10. Skills Check Percents are used with pie graphs. Express each as a percent.

a) 25 out of 100 people

b) 50 out of 200 light bulbs

c) 50 out of 500 cars

d) 50 out of 75 students

e) 2 out of 3 houses

f) 3 out of 5 dentists

g) 7 out of 10 drivers

h) 1 out of 4 telephones

11. In question 2, a bar graph shows where transportation workers are employed. Displaying only these seven areas of employment in a pie graph would not be meaningful. But you could display these seven areas and "Other" as percents of the total in a pie graph. Explain why you would need a section for "Other."

Transportation Workers: Where They Work

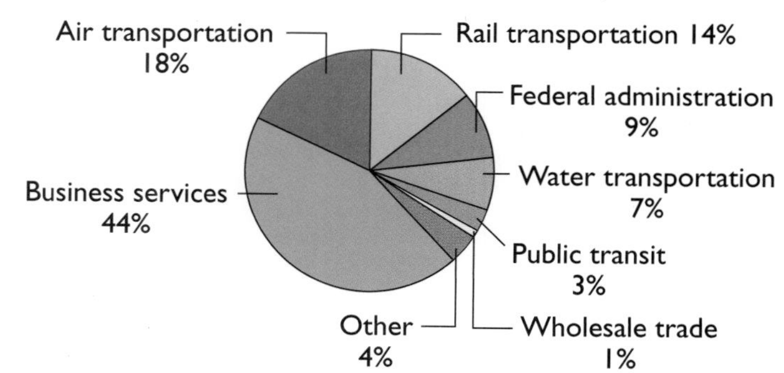

1.2 Constructing Graphs

Explore

Think of the types of graphs you saw in Section 1.1. How can you construct graphs like these?

Develop

1. Jamal does research for an environmental group. He interviewed 222 people about their methods of travel to work. The environmental group wants to collect statistics for a presentation to the government on improving public transportation.

	A	B
1	Method of Travel to Work	
2	Method of Travel	Number of People
3	Public transit	30
4	Motor vehicle driver	147
5	Motor vehicle passenger	19
6	Bicycle	4
7	Walk only	12
8	Not ascertained	10

AT

a) Construct a bar graph to display the results. Label the graph "Method of Travel to Work." Label one axis "Method of Travel" and the other axis "Number of People." Adjust sizes and colours until you are happy with the look of the graph.
b) Which is the most common method of travel to work?
c) Which is the least common method of travel to work?
d) Explain why a bar graph is suitable for displaying the data.
e) How might the environmental group use the data to request improvements to public transportation?

2. Susan found the American League East standings in the newspaper. She wants to compare the baseball teams' wins and losses.

AMERICAN LEAGUE EAST

TEAM	W	L
Boston Red Sox	43	28
New York Yankees	40	31
Baltimore Orioles	34	38
Toronto Blue Jays	34	38
Tampa Bay Devil Rays	21	51

AT

a) Construct a double bar graph to display the wins and losses for each team. Label the graph "American League East Standings." Label one axis "Team" and the other axis "Number of Games." Use a legend to show which bars represent wins and which represent losses. Adjust sizes and colours until you are happy with the look of the graph.

b) Determine the team with

i) the most wins **ii)** the fewest losses

iii) the fewest wins **iv)** the most losses

c) Explain how the answers in part b) are related.

d) Explain why a double bar graph is most suitable for displaying the data.

3. Kashmira works in a factory that plates copper onto printed circuit boards. She measures the concentration of copper in the chemical bath every week and compares the results. The spreadsheet shows the data for January and February.

	A	B
1	Concentration of Copper in a Chemical Bath	
2	Date	Concentration of Copper (g/L)
3	Jan. 4	15.06
4	Jan. 13	14.22
5	Jan. 20	13.86
6	Jan. 27	13.32
7	Feb. 3	12.90
8	Feb. 10	14.94
9	Feb. 17	14.10
10	Feb. 24	13.26

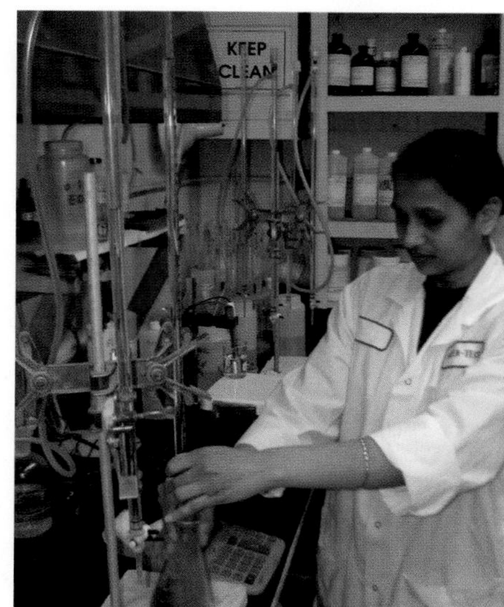

AT

a) Construct a line graph to display the data. Label the graph "Concentration of Copper in a Chemical Bath." Label one axis "Date" and the other axis "Concentration of Copper (g/L)."

b) When was the concentration decreasing?

c) When was the concentration increasing?

d) Explain why a line graph is suitable for displaying the data.

e) Would any other types of graphs be suitable for displaying the data? Explain.

Practise

4. **Skills Check** Find the difference between each high and low temperature.

   a) High: 5°C Low: 0°C
   b) High: 7°C Low: 1°C
   c) High: 3°C Low: −3°C
   d) High: 14°C Low: 2°C
   e) High: −6°C Low: −15°C
   f) High: 1°C Low: −9°C

5. The spreadsheet shows the average temperatures in Ottawa and Yellowknife in degrees Celsius.

	A		B
1	Average Temperatures in Ottawa and Yellowknife		
2	Month	Ottawa (°C)	Yellowknife (°C)
3	January	−10	−29
4	February	−10	−25
5	March	−5	−18
6	April	5	−8
7	May	15	1
8	June	20	11
9	July	25	18
10	August	20	11
11	September	15	8
12	October	9	0
13	November	0	−12
14	December	−8	−22

AT

Legend

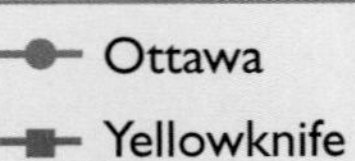

a) Construct a double line graph to display the data. Label the graph "Average Temperatures in Ottawa and Yellowknife." Label one axis "Month" and the other axis "Average Temperature." Use a **legend**, such as the one shown, to indicate what each line on the graph represents.

b) When and where was the temperature highest? lowest?

c) When was the greatest difference in temperature between the two cities? When was the least difference?

d) Is it easier to use the graph or the spreadsheet to answer part c)? Explain.

e) Explain why a double line graph is suitable for displaying the data.

6. Mike's company makes items to be sold at concerts. He summarizes the types and quantities of items that his company made last month.

	A	B
1	Concert Merchandise Produced Last Month	
2	Type	Number
3	T-shirts	1500
4	Hats	324
5	Headbands	260
6	Programs	850
7	Posters	235
8	Collector cards	140
9	Key chains	250

AT

a) Construct a pie graph to display the data.
b) Which type of merchandise is produced most frequently? least frequently?
c) Explain why a pie graph is suitable for displaying the data.
d) If you had data about monthly production from several months, the pie graph would no longer be suitable. Why?

7. What type of graph would you use to display each of the following types of data? Explain your choice.

a) the percents of your allowance that you use for savings, personal expenses, charity, and entertainment
b) the number of hours you spend on math homework each week for a term
c) average test scores in reading, writing, and math for your school
d) average test scores in reading, writing, and math for your school compared to the average test scores in those subjects for all schools in the province
e) the number of hours you spend on math homework each week for a term compared to the time you spend on English homework

8. If you were looking for an increase or a decrease over time, which type of graph(s) would be best for displaying the data?

1.3 Career Focus: Factory Worker

Trevor works in a small factory that makes printed circuit boards. The boards are used in computers.

Trevor puts blank boards in one end of a large machine. The machine plates copper circuits onto the boards. Trevor must read order forms to find the number of boards to make and how to make them.

Clearmount Circuit Boards

Customer:
Salem Computers Ltd.
1352 Bond Road, Unit 5
Our Town

Order No. 2037

Check boards for damage before starting to plate.

Part No. 53
Plating Thickness 0.05 mm
Plating Time 1 hour
Current 12 amps
Number Required 50

When completed, send boards to Inspection Department.

Date ____________

Operator's Signature ______________________________

1. a) What is Trevor's job?
b) How does he use reading skills in his job?
c) How many boards should Trevor plate for order No. 2037?
d) How thick should the copper plating be?
e) What must Trevor do before he puts a board in the machine?
f) What should Trevor do with the boards when he is finished?

2. Why is Trevor required to sign the form when he is finished processing the order?

3. Every week, the concentration of copper in the machine is measured and graphed. Trevor uses the graphs to keep the right concentration of copper in the machine. The concentration of copper must be kept between 12.00 g/L and 15.60 g/L. When it gets too low, Trevor adds more copper to the chemical bath.

   Trevor receives on-the-job training. He will soon be able to take the measurements and plot the graphs on his own.

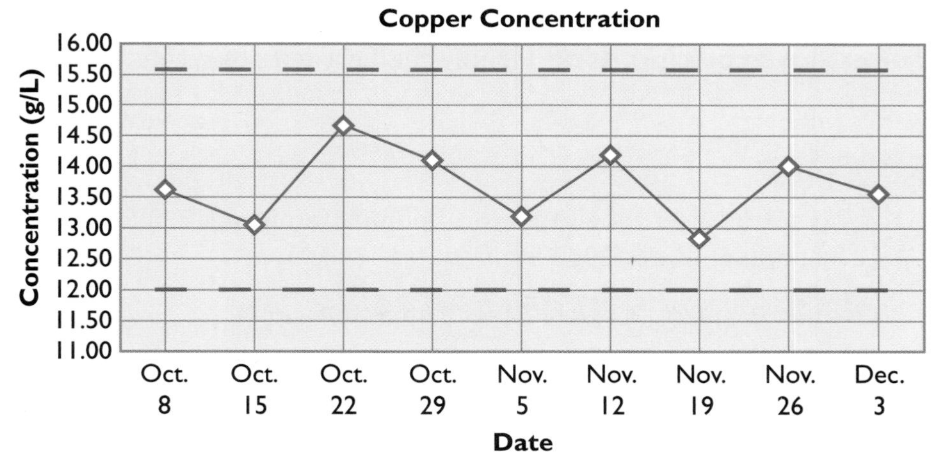

   a) How does Trevor use graphs in his job?
   b) What must Trevor learn so that he can do his job more effectively?

4. Use the graph above.
   a) When did the concentration of copper increase?
   b) When did the concentration of copper decrease?
   c) What changes did Trevor make to the concentration levels on or after October 15? Explain how you knew he made them based on the graph.
   d) On November 19, the concentration of copper was not below 12.00. Why do you think Trevor added more copper?
   e) Why do you think two horizontal dashed lines appear on the graph, one at 12.00 and the other at 15.60?
   f) Why is a line graph suitable in this situation?

5. Discuss the advantages and disadvantages of working in a small factory.

1.4 Constructing and Interpreting Graphs

There are many types of graphs. One type of graph may illustrate a specific type of data more effectively than another. Often, more than one type of graph can display the data effectively.

In this section, you will create the type of graph that you think best illustrates the data given.

Explore

You want to show that you are trying to improve your marks. What types of information could you collect to demonstrate your efforts? For each type of information, identify the best type of graph.

Develop

1. This spreadsheet shows the world population from 1750 to 2050. The population for 2050 is, of course, an estimate.

	A	B
1	World Population	
2	Year	Population in Millions
3	1750	791
4	1800	978
5	1850	1262
6	1900	1650
7	1950	1521
8	2000	6073
9	2050	8909

AT

a) Use the spreadsheet to construct graphs. Experiment with various types of graphs to determine how best to display the data.

b) Identify each type of graph you tried for part a). Discuss the pros and cons of each type.

c) Which type of graph do you think is most suitable for the data? Why?
d) In what time interval was the greatest increase in world population?
e) Why do you think a greater increase occurred at this time compared to other times?
f) How do you think that the world population estimate for 2050 was determined?

2. This set of data shows the populations of Europe and North America from 1750 to 2050.

	A	B	C
1	Population in Europe and North America in Millions		
2	Year	Europe	North America
3	1750	163	2
4	1800	203	7
5	1850	276	26
6	1900	408	82
7	1950	547	172
8	2000	729	305
9	2050	628	392

AT

a) Use the spreadsheet to construct graphs to display the data. Experiment with types of graphs until you determine the graph that you think best displays the data.
b) Why did you decide upon the type of graph you chose in part a)?
c) In what time interval was the greatest increase in the population of Europe?
d) In what time interval was the greatest increase in the population of North America?
e) Why do you think the population of North America rose dramatically between 1850 and 1900?
f) Why do you think it is predicted that the population of Europe will decrease between 2000 and 2050?

Practise

3. This set of data shows the energy burned by a 68 kg person during various activities.

	A	B
1	Energy Burned When You Exercise	
2	Activity	Energy Burned (kJ/min)
3	Aerobics	29
4	Cycling (leisurely)	29
5	Cycling (moderate)	38
6	Basketball	38
7	Jogging	38
8	Running up stairs	71
9	Soccer	33
10	Swimming	38
11	Tennis	33
12	Bowling	12

a) Use the spreadsheet to construct a graph to display the data. Experiment with types of graphs until you determine the graph that you think best displays the data.

Access to the Web site for calculating energy burned can be gained through the *Mathematics for Everyday Life 12* page of www.math.nelson.com.

b) Why did you decide upon the type of graph you chose in part a)?

c) Select one of the other types of graphs. Explain why you think it did not show the data well.

d) Which activity burns the least energy?

e) If you were interested in burning more energy, which two activities from this list would you choose? Explain.

f) Why are people interested in burning energy?

4. **Skills Check** Estimate.

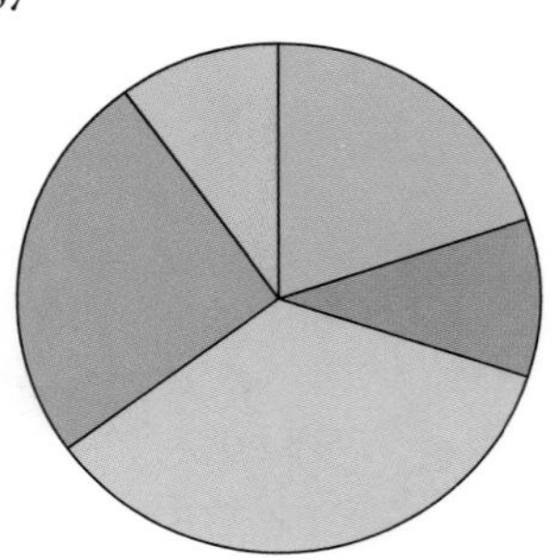

a) About what percent of the graph does the red sector represent?

b) Which two sectors represent about the same amount?

c) Which sector represents about 35%?

d) Which sector represents about 20%?

5. The spreadsheet below shows the immigrant populations of Hamilton and London, Ontario, by place of birth. Statistics Canada collected this information in the 1996 **census**. A census is an official count of the population and a statistical record of other things of interest to researchers.

	A	B	C
1	Immigrant Populations by Place of Birth		
2	Place of Birth	Hamilton	London
3	United States	5360	4045
4	Central and South America	5225	4020
5	Caribbean and Bermuda	5045	2075
6	Europe	105105	49320
7	Africa	2985	2665
8	Asia	21335	13340
9	Oceania	600	515

AT

a) Use the spreadsheet to construct the best graph to display the data.
b) Why did you choose the type of graph you constructed in part a)?
c) For both Hamilton and London, where was most of the immigrant population born?
d) What other observations can you make about the data?

6. This spreadsheet shows how Stephan budgets his money.

	A	B
1	Stephan's Monthly Budget	
2	Category	Amount ($)
3	Entertainment	75
4	Transportation	22
5	Clothing	22
6	Food	37
7	Savings	7
8	Other	14

AT

a) Use the spreadsheet to construct a graph.
b) Why did you choose the type of graph you constructed in part a)?
c) On what budget category does Stephan spend most of his money?
d) On what budget category does Stephan spend the least money?
e) How could Stephan alter his budget if he decides to save money to buy a car? Explain your suggestions.

1.5 Misleading Graphs

Most of the graphs that you have seen in this chapter represent data without misleading the viewer. On page 5, however, you examined two Distribution by Age graphs where the age groups were not the same size. You might question
- what the graph would look like if the age groups had been equal, perhaps by decade
- whether the tallest would be the tallest if its age group wasn't such a large age group
- why the age groups used were chosen

Graphs can create misleading impressions. Often this is intentional. Sometimes, it is not.

Explore

What impression is created by each graph?
Which graph is not misleading? Why?
Why is the other graph misleading?
Why might the misleading graph be used?

a)

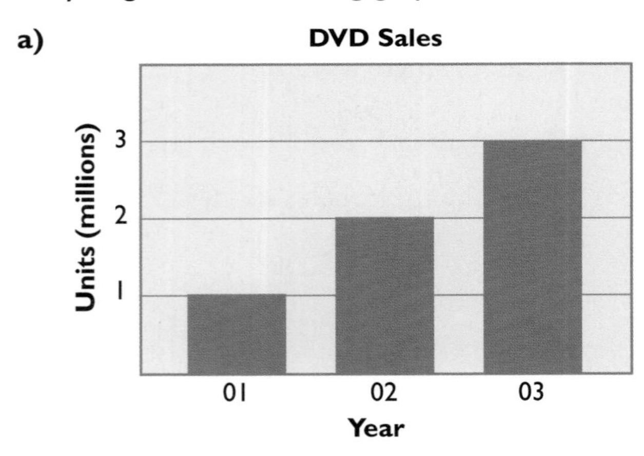

b)

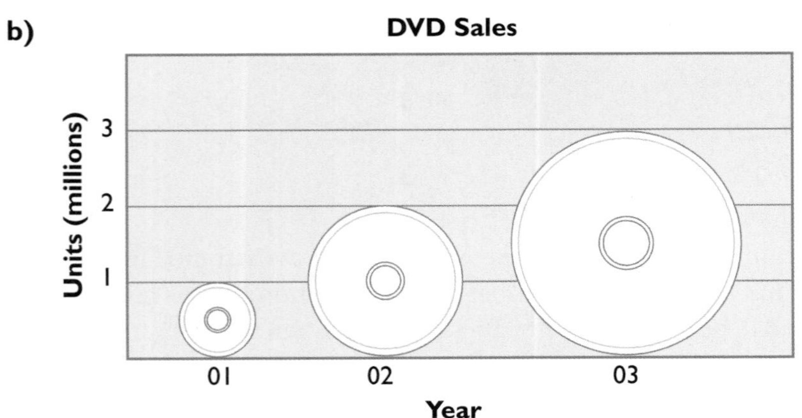

Develop

1. a) Which graph displays the data without misleading?

i)

House Prices

212 000
210 000
208 000
206 000
204 000
202 000
200 000
198 000
196 000
194 000

Average House Price ($)

2002
2003

Year

ii)

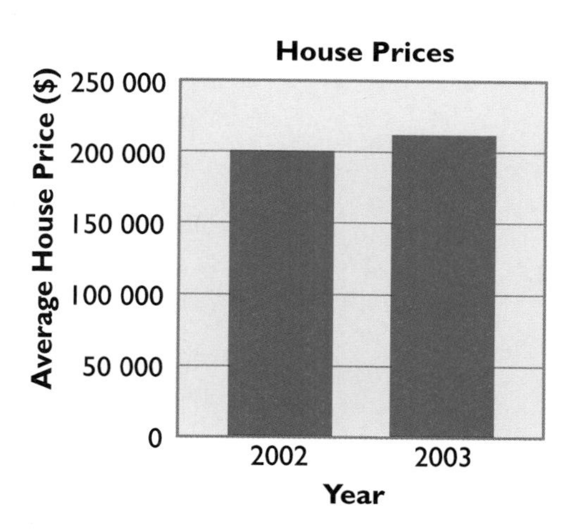

b) What about the construction of the other graph creates the misleading impression?

c) What is the percent increase? In other words, what percent of the 2002 price, $200 000, is the increase, $10 250?

d) The percent increase in part c) is significant, even more so if earnings and other prices have not increased as much. Why is the following graph a less misleading way to display data where values are large numbers and starting a scale at 0 shows little change?

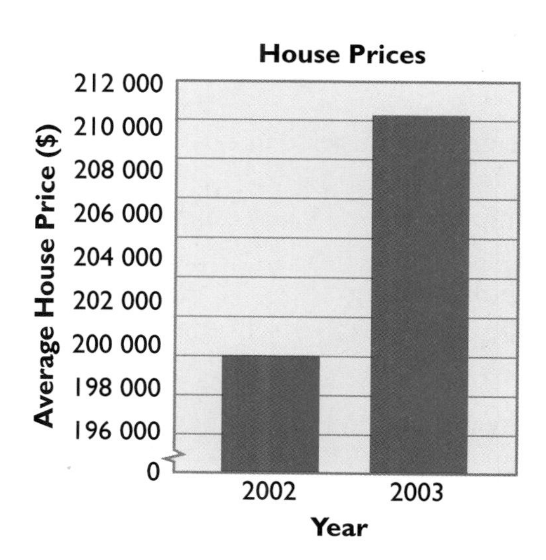

Practise

2. a) How is the graph misleading?

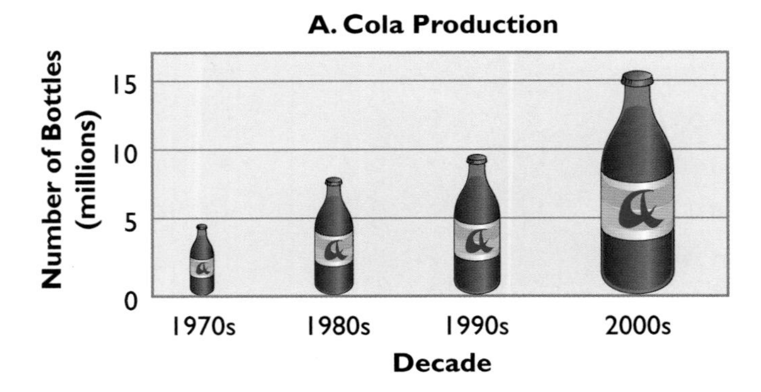

b) Who might use the graph and why?
c) What changes would make the graph more accurate and less misleading?

3. A company shows the graph below to its shareholders and customers.

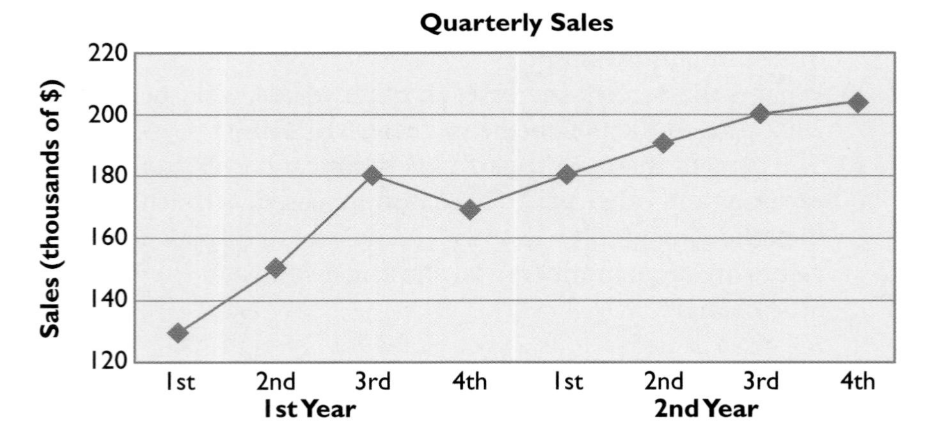

a) Which quarter had the greatest increase in sales over the previous quarter?
b) Describe the sales during the second year compared to those in the first year.
c) What makes this graph misleading?
d) How could the graph be changed to present the data more accurately?

AT 4. Display the data in the spreadsheet on the next page in a graph each of the following ways. Explain each graph.
a) the way the directors might if they wanted to minimize their expenses
b) the way the news media might if they wanted to create an impression of overspending on the part of the directors

	A	B
1	AD CO Directors Expense	
2	Quarter	Amount ($)
3	1st	92000
4	2nd	109000
5	3rd	128000
6	4th	164000

5. a) What impression is created by each graph?

i)

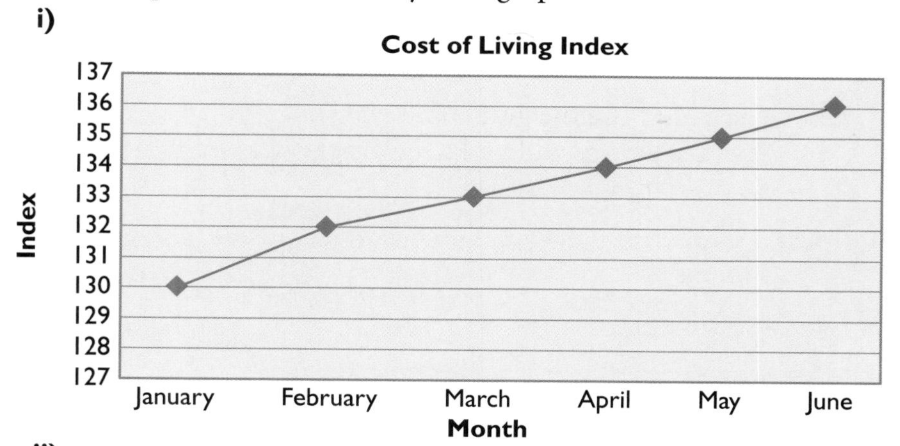

ii)

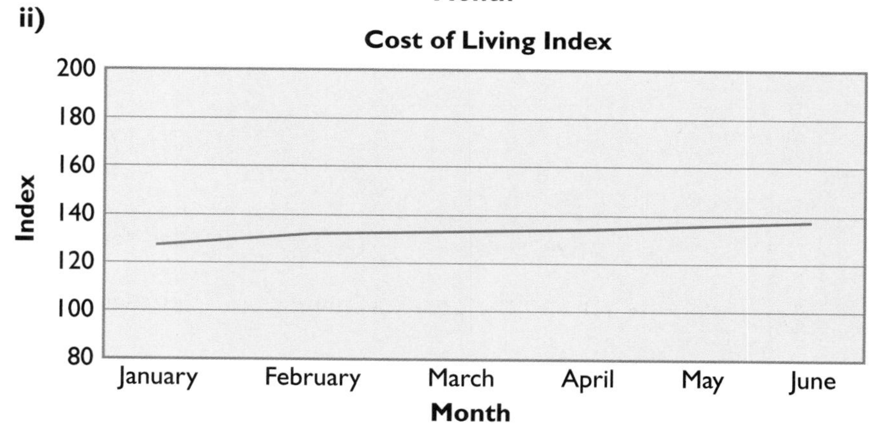

b) Research the cost of living index and suggest a way that the graphs could display the data more accurately and without being misleading.

Access to Web sites about the cost of living index can be gained through the *Mathematics for Everyday Life 12* page of www.math.nelson.com.

6. Look for graphs in newspapers, magazines, and on the Internet. Find at least three that mislead. Discuss how each graph creates a misleading impression. Describe how each could be changed to display the data more accurately and without being misleading.

1.6 Putting It All Together: Selling Shoes

Amela manages a shoe store. These spreadsheets show data that Amela collected about her sales.

	A	B
1	Yearly Sales of Shoes	
2	Year	Sales ($)
3	1995	351000
4	1996	329000
5	1997	367000
6	1998	382000
7	1999	410000
8	2000	415000
9	2001	435000

	A	B
1	Sizes of Shoes Sold in September	
2	Size	Number of Pairs Sold
3	6	70
4	7	245
5	8	221
6	9	148
7	10	53
8	11	31
9	12	32

	A	B
1	September Sales by Category	
2	Type of Shoe	Sales ($)
3	Women's Dress	11500
4	Women's Casual	14200
5	Men's Dress	5100
6	Men's Casual	9200

AT **1. a)** Use the spreadsheets to construct three graphs to represent the data without misleading.

b) For each graph you constructed for part a), explain why you decided upon that type of graph.

Use your graphs from question 1 to answer questions 2 to 6.

2. What happened to Amela's yearly sales between 1995 and 1996?

3. What would you expect Amela's yearly sales to be in 2002? Why?

4. What size of shoe was sold most frequently in September?

5. In September, Amela sold 800 pairs of shoes. She will soon be placing an order for 400 pairs of shoes. About how many pairs of shoes should she order in size 9? Explain.

6. Which type of shoe was sold the most in September?

7. How could you change the graph for the first set of data to give the impression that sales were increasing greatly in recent years?

8. What about the graphs suggests that Amela sells more women's shoes than men's shoes?

1.7 Chapter Review

1. Give examples of situations for which each type of graph could be used.

a) bar graph
b) double bar graph
c) line graph
d) double line graph
e) pie graph

2. Misha receives a property tax bill for $2728. A pie graph shows how his property taxes are spent.

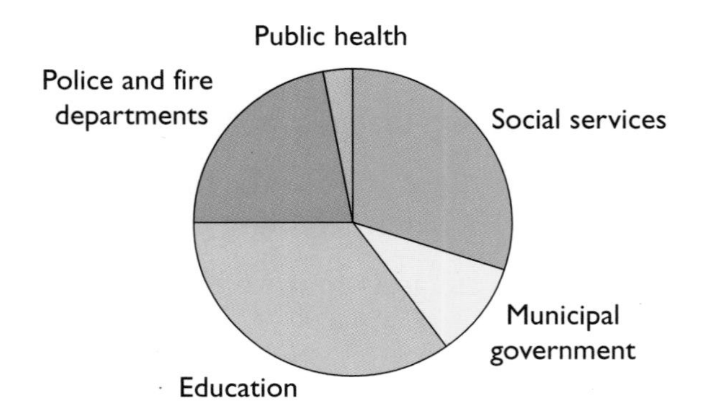

a) Why do you think a pie graph is used to represent the data?
b) To what area do most of Misha's property tax dollars go?
c) What area receives the least amount of his property tax dollars?
d) Identify two areas that together take more than half of Misha's property tax dollars.

3. A report released by Statistics Canada shows that as Canadians spend more time on the Internet, they spend less time with family and friends.

	A	B	C
1	Internet Use		
2	Time Spent on Internet per Week (hours)	Percent of Users Spending Less Time with Family per Week	Percent of Users Spending Less Time with Friends per Week
3	less than 5	5	4
4	5 to 15	8	8
5	more than 15	14	13

a) Use the spreadsheet to construct a graph to accurately represent the data.
b) Which group spent less time with friends?
c) Examine the data for users who spend 5 to 15 hours per week on the Internet. What do you notice about these figures compared to the data for the other two groups?
d) Explain why a double bar graph is suitable for displaying the data.
e) Which type of graph would not be suitable for displaying these data? Why?

4. Construct a graph to display the data below. Explain why you chose that type of graph. Why did you decide against the other types?

	A	B
1	Michael's Height	
2	Age (years)	Height (cm)
3	1	80
4	2	91
5	3	101
6	4	108
7	5	116
8	6	125

5. Use your graph from question 4.
a) Was there any year in which Michael grew faster than in other years? How do you recognize this in the graph?
b) How could you estimate Michael's height at age seven?

6. a) What impression is created by this graph?
b) Is this a misleading impression? Explain.

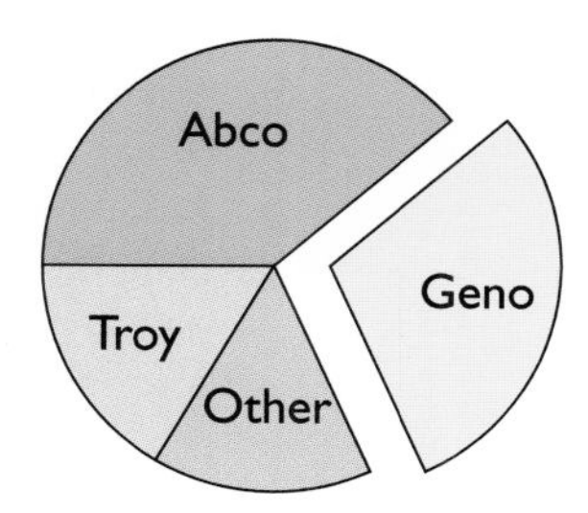

7. A new company has been very successful. Sales this year were double the previous year's sales of $1.5 million. Describe two ways you could graph the data to exaggerate this increase. Do you consider the ways you suggested to be misleading? Explain.

CHAPTER

2 Collecting and Organizing Data

In this chapter, you will
- use sampling techniques
- interpret and assess statistics in the media
- collect and record data
- construct tables and graphs to represent data
- draw conclusions by interpreting graphs

Near the end of the chapter, you will collect data to study an issue that interests you.

2.1 Sampling Techniques

People collect data on a wide variety of topics for various uses. Politicians and marketing firms, for example, collect data on people's opinions. Manufacturers collect data for quality control. The data collected help people make decisions.

Explore

This question appeared on a survey:

> Do you think the use of pesticides should be allowed for killing weeds on lawns?
>
> Yes   No   No opinion

What types of organizations might ask this question? Do you think it is a good survey question? Why?

Develop

Collecting data from an entire **population** is usually costly and impractical. The population is the whole group being studied. Data are usually collected from a **sample**, a small group from the population. For the data to be useful, the sample must represent the population.

An environmental group wants to gather data about water pollution in Ontario. Testing just one lake, such as Lake Ontario, would not adequately represent the population of bodies of water in Ontario. The sample to be tested should include lakes and rivers of various sizes from various water systems throughout the province.

1. For each population identified below, discuss why the proposed sample would not be representative of the population.

a) To find the number of cars owned per family in Canada, collect data in Edmonton and Quebec City.

b) To determine the opinions of Canadians about offshore fishing, collect data in communities in Newfoundland.

c) To check the quality of cars produced in a plant, test the cars produced during the second shift on Tuesday.

d) To collect data about the most popular fast-food restaurant in a city, ask people arriving at one of the city's fast-food restaurants.

Sampling techniques are the methods used to gather data from a population. The techniques used depend upon the data to be gathered and the population of interest. They include

- surveys
  - by telephone
  - by personal interview
  - by mail
- experiments
- counting or measuring

2. Which sampling technique was likely used to gather data in Chapter 1 on

a) amount of snowfall? (page 3)
b) ages of people working in transportation occupations? (page 5)
c) method of travel to work? (page 10)
d) concentration of copper in a chemical bath? (page 11)

3. Surveys must be simple and clear. The questions should be written so that they don't produce a biased response.

Discuss each survey question below. Improve the wording where needed.

a) There has been an increase in the incidence of vandalism. Do you think more money should be spent on security?

Yes   No   No opinion

b) How would you rank the Prime Minister's performance in the past six months?

Good   Fair   Poor   No opinion

Practise

4. Which sampling technique would you use to determine each of the following? Why?

a) the popularity of a new TV series
b) the concentration of a chemical in water
c) the need for traffic lights at an intersection
d) the quality of fireworks being produced
e) the opinions of voters on crime

5. Examine each survey question. Improve the wording where needed.

a) Given the recent chemical spill, do you think that the trains should be re-routed through less densely populated areas?

Yes No No opinion

b) Are you in favour of a toll highway between Peterborough and Ottawa?

Yes No No opinion

c) Which brand of soft drink tastes the best?

Tasty Treat Sensational Raspberry Delight Drink-Up Red Crush

d) Dolphins should be kept in captivity to attract tourists.

Agree   Disagree

5 4 3 2 1

6. Suppose you need to determine the popularity among adults in Ontario of the following sports on TV—hockey, football, basketball, racing, golf, and skiing.

a) Choose a sample for the population and describe it.
b) Choose a sampling technique and explain why it is suitable.

7. Merle works in quality control for a manufacturer of breakfast cereals. What sampling techniques might Merle use for quality control? Why would the manufacturer want to do this?

8. Suppose you work for a painting company. A lot of your business is in exterior painting and depends on the temperature outside. You need to gather temperature data for the area to determine your peak painting season.

a) What sampling technique would you use to gather the data?
b) How would you record the data?

9. Create a biased and an unbiased survey question on each topic.

a) the popularity of a new TV series
b) the opinion of residents on crime in their area

2.2 Statistics in the Media

Explore

Consider the advertising slogan below. What does the slogan actually tell you? What doesn't it tell you? What would you like to know to help you decide whether or not to buy Kleenteeth?

"Seven out of ten dentists recommend ***Kleenteeth***"

Develop

As well as in advertising, the media regularly report statistics in the news.

About 60 percent of Canadians believe their weight is ideal: poll

Environment getting worse, Canadians believe

July jobless rate unchanged at 7%
Ontario rate rises from 6% to 6.3%

1. The headlines above might prompt questions such as these:
- What is meant by ideal weight (or environment or jobless rate)?
- Is the statement true for my friends and family members?
- Where did the information come from?
- If the information is from a survey, who was surveyed? By whom?

What other questions do these headlines raise?

Practise

2. Read the following news clipping. Answer the questions below it to help you interpret and assess the information.

Poverty top children's issue

If Canadian children could change one thing affecting the lives of young people in their country, it would be to address poverty, a report to be released today reveals.

The report, commissioned by the federal government to guide Canada's input at an upcoming United Nations meeting on children, found poverty topped the list of concerns for 18.5 per cent of 1,200 children surveyed.

"Some of (the children) were recounting their own personal situations of poverty, but they were also looking outside of themselves and saying it's unacceptable that people are living on the street and that children don't have enough to eat," said Alana Kapell, who compiled the report for the non-profit organization Save the Children Canada.

The respondents' definitions of poverty were broad, citing any barriers to eating nutritious food, attending school, sleeping in a bed and having clothes to wear.

Kapell's report, called "A Canada Fit for Children," surveyed children aged 7 to 18 in 57 communities across Canada. …

Other big concerns for Canadian youth are abuse and violence, listed as the No. 1 concern for 13.5 per cent of children surveyed, and drugs, alcohol or smoking, listed as No. 1 by 12.2 per cent of respondents. More than half of the survey respondents said they wanted adults to "listen to them, understand, believe in them, and remember that it is difficult being young."

Excerpt from *Poverty top children's issue* by Nancy Carr, Canadian Press, August 14, 2001

a) Who conducted this survey?
b) For whom was it conducted?
c) By whom was the survey funded?
d) What questions might have been asked in this survey?
e) Describe the sample and population for this survey.
f) What is the sample size?
g) What is the main concern for children and youth today?
h) What is the definition of poverty according to the children and youth of today?
i) What questions could you ask to help you interpret the results of the survey?
j) How well do you think the headline represents the data? Explain.

3. Locate a news article in a newspaper, magazine, or on the Internet that discusses the results of a survey or study. Answer parts a) to f) and i) and j) from question 2.

4. Sometimes, statistics are misinterpreted or misused by accident or lack of knowledge. Sometimes, misuse is deliberate, with the intention to mislead or to create an attraction or sensation. Locate statements in headlines that appear to be misusing statistics to create an attraction or sensation. Explain why you consider them to be misusing statistics.

2.3 Organizing and Interpreting Data

Explore

What does this graph tell you? What doesn't it tell you? What would you like to ask about the survey or the graph to help you get a better picture of the situation?

Radio Listening Time of Canadians Aged 12 to 17

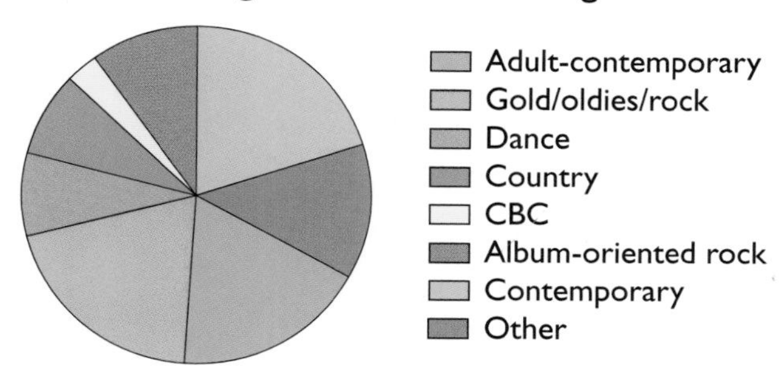

Source of data: Statistics Canada

Develop

1. A **tally chart**, or **frequency table**, is a useful method for organizing data collected. The possible responses are listed and a tally mark is recorded beside a response each time it is given. Then the tally marks for each response are totalled and all the responses are totalled.

The table shows data collected about the popularity of Canadian hockey teams among secondary school students in Toronto.

Hockey Team	Tally	Frequency
Edmonton	~~IIII~~	5
Toronto	~~IIII~~ ~~IIII~~ I	11
Montreal	~~IIII~~ I	6
Calgary	III	3
Vancouver	IIII	4
Ottawa	~~IIII~~ I	6
Total		35

The data can be displayed using various types of graphs.

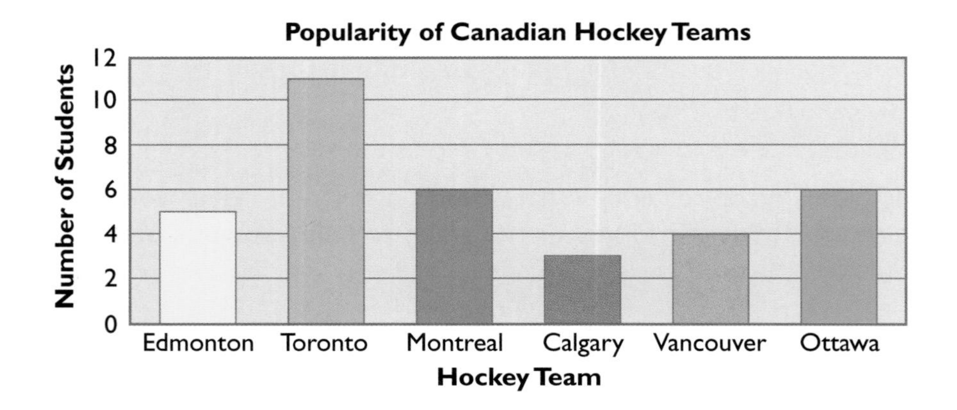

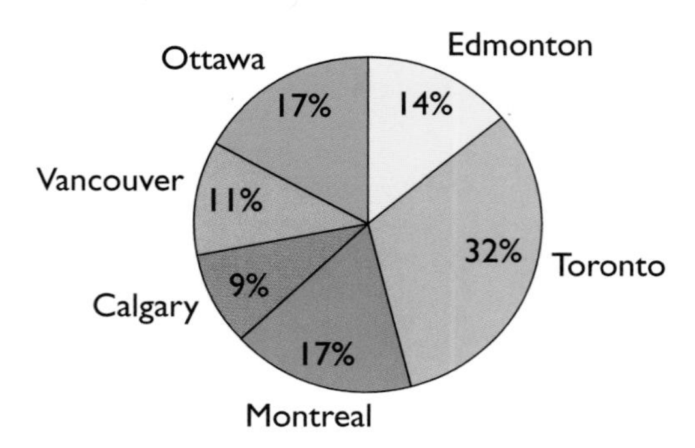

a) Describe the population and the sample used.
b) Which graph displays the data most effectively?
c) Which team is the most popular among the sample? the least popular?
d) How useful are the data from the sample for making conclusions about the population?
e) Who might use this information? How?

Practise

2. a) List five TV shows on Tuesdays during prime time (8 p.m. to 11 p.m.).
   b) Write a question that you could ask your classmates to gather some data about the shows.
   c) Create a tally sheet to record the data collected.
   d) Carry out the survey of your classmates.

3. Nico collected the following data. She presented names of five shows on TV on Tuesdays during prime time. The survey question was "Which of these shows is your favourite?"

TV Show	Tally	Frequency
Laugh At This	𝍸	5
PVFD	𝍸 𝍸 II	12
City General	I	1
Trivia Time	III	3
The House	𝍸 II	7

   a) Construct a graph using the data.
   b) If you were a TV producer, which show would you take off the air?
   c) If you were an advertiser, which show would you advertise with? Why?
   d) Who else might use information such as this?

4. a) List five popular fast foods.
   b) Write a question that you could ask your classmates to gather some data about the foods.
   c) Create a tally sheet to record the data collected.
   d) Carry out the survey of your classmates.

5. a) List five items on which your classmates spend money, such as going to a movie.
   b) Write a question that you could ask your classmates to gather some data about their spending.
   c) Create a tally sheet to record the data collected.
   d) Carry out the survey of your classmates.

6. This set of data is based on 100 words in a book.

Number of Letters	Tally	Frequency
1	III	3
2	~~IIII~~ ~~IIII~~ ~~IIII~~ IIII	19
3	~~IIII~~ ~~IIII~~ ~~IIII~~ ~~IIII~~ III	23
4	~~IIII~~ ~~IIII~~ ~~IIII~~ ~~IIII~~	20
5	~~IIII~~ III	8
6	~~IIII~~ ~~IIII~~	10
7	~~IIII~~ II	7
8	III	3
9	III	3
10 or more	IIII	4

AT

a) Construct a graph.
b) Why did you choose the type of graph you constructed in part a)?
c) What length of word is most common? least common?
d) Is it easier to find the answer to part c) from the table or the graph? Why?
e) What type of graph would you choose to display the data from this question and from a count of the lengths of words in a children's book?
f) What do you think you would find in comparing the data using the graph from part e)?
g) Who might use this type of data? For what purpose?

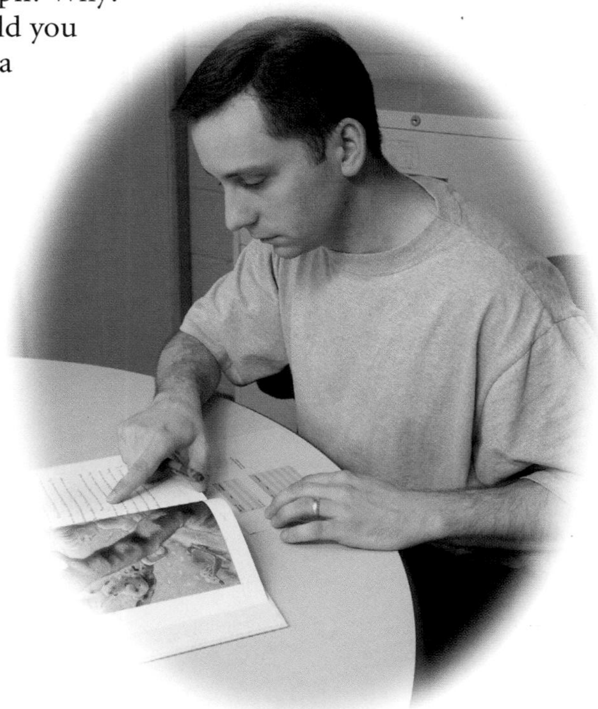

2.4 Career Focus: Telemarketer

Allan works as a telemarketing representative. He makes sales over the phone and keeps records of interested customers and sales.

Other jobs in telemarketing include
- conducting polls or surveys over the phone
- raising funds for charitable organizations over the phone

Allan got the job because he has these skills and qualifications:
- good communication skills in English
- ability to speak clearly and hear well over the phone
- good customer service skills
- ability to sit for reasonably long periods
- keyboarding and data entry skills
- knowledge of sales techniques
- a high school graduation diploma

Allan works at a call centre. He works on his own using a headset and recording the information from calls on a computer. He usually works evenings and weekends. His employer provides on-the-job training.

1. a) What are Allan's main tasks as a telemarketer?

b) How does Allan use a computer in his job?

c) How does Allan train to be a telemarketer?

d) Why do you think telemarketers work mostly evenings and weekends?

2. Think about the types of telemarketing calls you have received or heard about. Make a list of types of surveys, sales, and fundraising that might be done over the phone.

3. A TV station wants to determine the number of Canadian TV viewers who watch a new TV series compared to the number who watch other shows of the same type.

a) How might a telemarketing company select a sample from the population of Canadian TV viewers?

b) What question might the company have its representatives ask to gather the information?

c) How might telemarketers record the data?

d) What type of graph would be appropriate for displaying the data?

e) How might the data be used?

4. Suppose you want to determine the popularity of fast-food restaurants in your area.

a) Name five fast-food restaurants in your area.

b) Write a brief description of the types of food served at each restaurant.

c) Write one or more survey questions that could be used by a telemarketer to collect data about the popularity of fast-food restaurants.

d) Make up a tally sheet to record the data.

e) What type of graph could display the data well?

f) If you conducted the survey, who might use the results? How would they use the survey results to make decisions?

2.5 Putting It All Together: Conducting a Survey

1. a) Identify an issue or topic of interest to you, about which you would like to collect some data.
   b) Write two or more survey questions for your issue or topic.
   c) Explain why your survey questions are suitable.
   d) List four to six possible responses for each question.
   e) Explain why you developed those responses for part d).
   f) Decide on a population and the best sample for your survey. Describe the sample.
   g) Describe a method for conducting the survey and explain why it is a suitable method.
   h) Predict what you think the results would show.

2. Conduct the survey from question 1 using your class as the sample. Use a tally sheet to record the data.

3. Use the data collected from your survey in question 2.
   a) Enter the data in a spreadsheet.
   b) Construct a bar graph, a line graph, and a pie graph. (AT)
   c) For each graph, explain why it is or is not an effective way to display the data.
   d) From your data and graphs, draw a conclusion or make a general statement to answer the survey questions.
   e) Compare the results with your prediction from question 1 g). Give reasons for any differences.

2.6 Chapter Review

1. Martha works for a polling company and needs to study shoppers' buying patterns at grocery stores in Barrie.

a) What population does she want to study?
b) Describe the sample she might use for a survey by mail.

2. Gregory is setting up a concession stand at the community centre. He wonders what types of food he should sell.

a) Describe the population he would be interested in.
b) Describe a sampling technique that Gregory could use.
c) Write a survey question for him to ask.

3. Discuss whether the sample and sampling technique outlined below were suitable or not. If not, make recommendations for improvements.

To determine whether smoking is healthy or not, a pollster interviewed randomly selected smokers outside office buildings in downtown Toronto.

4. List at least five questions that this headline raises.

Four out of 5 Canadians not satisfied with health care

5. Tien surveyed people in her community about the amount of time they spent on various physical activities. She organized and graphed the results.

a) Draw conclusions about what the graphs show.
b) What additional information would you like to know about the survey or the graphs?
c) Which graph is most suitable for the data? Which graph is not suitable? Explain.

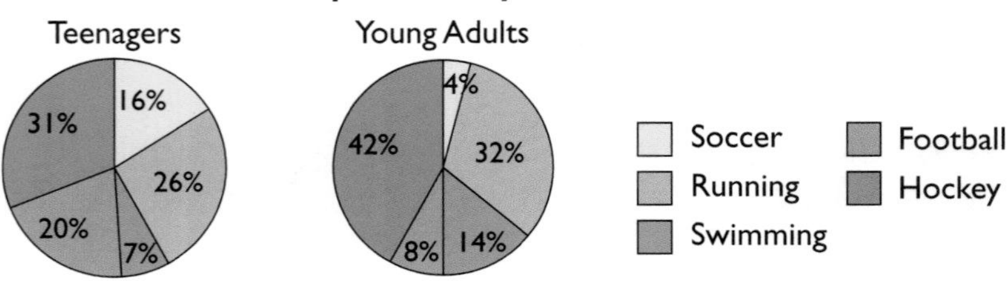

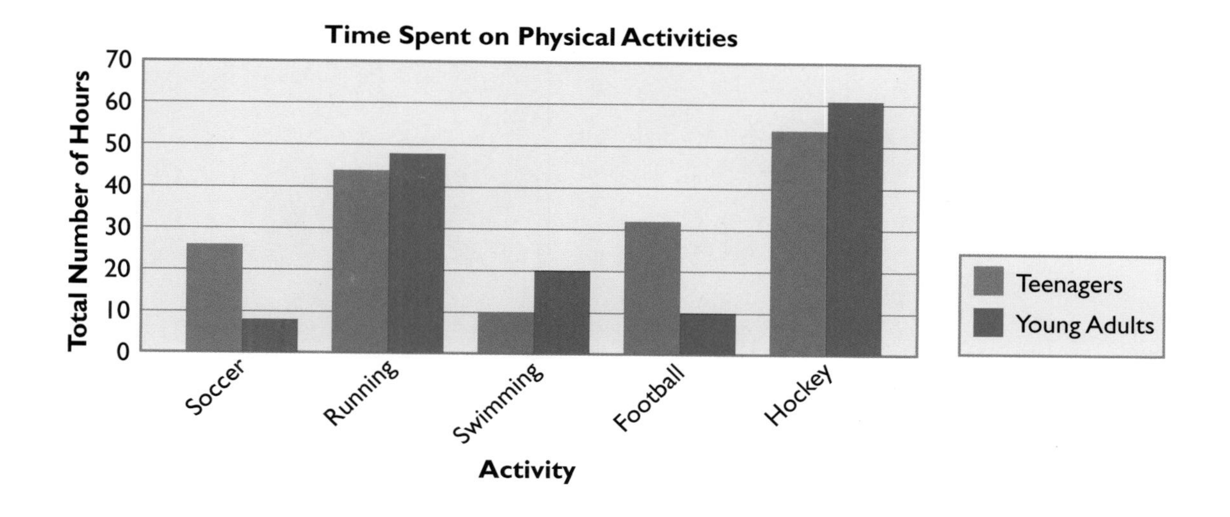

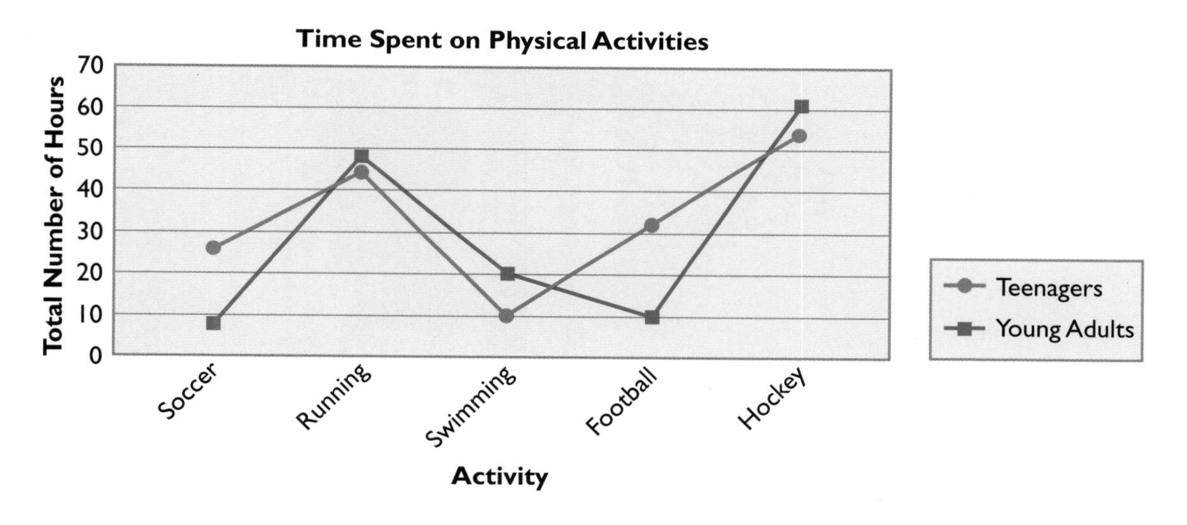

6. Miguel conducted a survey to predict who would win the student election at his school. He randomly selected 25 students from various grades. The results for the five candidates are as follows:

Ahmed, Kathy, Carlos, Fatemeh, Kathy, Beth, Ahmed, Fatemeh, Beth, Fatemeh, Ahmed, Beth, Carlos, Fatemeh, Ahmed, Fatemeh, Ahmed, Fatemeh, Beth, Ahmed, Fatemeh, Beth, Fatemeh, Beth, Fatemeh

a) Construct a tally sheet.

AT **b)** Construct a bar graph, a line graph, and a circle graph.

c) Which graph displays the data most effectively? Explain.

d) Draw conclusions from the data and graphs.

CHAPTER

3 Probability

In this chapter, you will

- interpret experimental probabilities to make predictions and help make decisions
- calculate theoretical probabilities of simple events
- express probabilities as fractions, decimals, and percents
- perform probability experiments and compare the results with theoretical probabilities
- simulate familiar situations involving probability

Near the end of the chapter, you will

- calculate the theoretical probability of a simple event
- collect experimental data to support the theoretical probability of a familiar situation
- simulate the familiar situation to support the theoretical probability

3.1 Making Predictions

The probability of precipitation on a particular day is 30%. The probability of winning a particular lottery is 1 in 1000. The probability of being struck by lightning is 1 in 500 000. Probabilities, like statistics, are all around us. In simple terms, **probability** is the likelihood or chance of something happening.

Probability is a branch of statistics. In **empirical**, or **experimental, probability**, data are collected to make predictions.

Explore

Cameron was at bat 297 times and hit safely 102 times. Ted was at bat 248 times and hit safely 88 times. For which player is the experimental probability greater for making a safe hit the next time he is at bat? Explain.

Develop

1. The date the ice broke up on the lake where Chris's family has an island cottage has been recorded since 1901. It has always been between March 20 and May 5. This tally chart shows the number of times it broke up on each date over the 100 year span of 1901 to 2000.

<table>
<tr><td>Mar. 20 |</td><td>21</td><td>22</td><td>23 ||</td><td>24</td><td>25 |</td><td>26 ||</td></tr>
<tr><td>27 |</td><td>28 |</td><td>29 ||</td><td>30 ||</td><td>31 |</td><td>Apr. 1</td><td>2 |</td></tr>
<tr><td>3 |</td><td>4 |||</td><td>5 |</td><td>6 |</td><td>7 ||</td><td>8 ||</td><td>9 |</td></tr>
<tr><td>10 |</td><td>11 |</td><td>12 |</td><td>13 ||</td><td>14 |||</td><td>15 |||</td><td>16 𝍸</td></tr>
<tr><td>17 𝍸 ||</td><td>18 𝍸 |||</td><td>19 𝍸 ||</td><td>20 𝍸</td><td>21 |||</td><td>22 ||</td><td>23 ||</td></tr>
<tr><td>24 ||</td><td>25 𝍸</td><td>26 ||||</td><td>27 ||</td><td>28 |</td><td>29 ||</td><td>30 |</td></tr>
<tr><td>May 1 ||||</td><td>2</td><td>3 ||</td><td>4 |</td><td>5 |</td><td></td><td></td></tr>
</table>

The experimental probability of the ice breaking up on particular date(s) is $\frac{\text{the number of times on the date(s)}}{\text{the number of years studied}}$.

Determine the probability, expressed as a fraction, of the ice breaking up

a) on March 20
b) on April 25
c) on May 12
d) on March 10
e) in the month of March
f) in the month of April
g) in the month of May
h) before the beginning of May
i) before the beginning of April
j) between March 20 and May 5

Use the probabilities you determined in question 1 to answer questions 2 and 3. Explain your answers.

2. **a)** Would you believe that the ice broke up on June 1 in 2001?
   b) Would you believe that the ice broke up on March 15 in 2002?
   c) Which is the most probable date from question 1 parts a), b), and c) that the ice broke up on in 2003?
   d) How might Chris's family use this information?

3. **a)** What do you notice about the answers to question 1 parts e) and i)?
   b) What do you notice about the answers to question 1 parts c) and d)? What is something in your everyday life that has the same probability?
   c) What do you notice about the answer to question 1 part j)? What is something in your everyday life that has the same probability?
   d) In which month is the probability the greatest that the ice will break up?

Practise

4. **Skills Check** Use your knowledge of decimals to list the batting averages from the most probable to make a safe hit next time at bat to the least probable.
   0.284 0.373 0.432 0.401 0.337 0.291 0.410

5. A fitness study asked 300 people between the ages of 15 and 74 if they exercised at least three times a week. The yes responses were tallied by age groupings.

Age	Tally	Age	Tally
15–24	𝍸 𝍸 𝍸 𝍷𝍷𝍷	45–54	𝍸 𝍸 𝍸 𝍸 𝍸 𝍸 𝍸 𝍸 𝍸 𝍷𝍷
25–34	𝍸 𝍸 𝍸 𝍸 𝍸 𝍸 𝍸 𝍸 𝍸 𝍸 𝍸 𝍸 𝍷𝍷	55–64	𝍸 𝍸 𝍸 𝍸 𝍷
35–44	𝍸 𝍸 𝍸 𝍸 𝍸 𝍸 𝍸 𝍸 𝍷𝍷𝍷𝍷	65–74	𝍸 𝍸 𝍸 𝍷𝍷𝍷

Determine the experimental probability, expressed as a fraction, that the response would be yes if the same question was asked of one more person

a) aged 15 to 24 **b)** aged 35 to 44
c) over 24 **d)** under 45

6. How would the information in question 5 be used by each of the following?
   a) a government agency advertising the importance of fitness
   b) a fitness club advertising their facilities

3.2 Making Decisions

If the weather report indicates that the probability of precipitation is 70% in the afternoon, you might decide to take an umbrella with you when you go out in the morning for the day.

Many decisions are made based on probabilities. For example, insurance premiums, the amount an insurance company charges for insurance coverage, are determined by collecting data. Premiums vary depending upon the probability of a claim being made. The greater the risk to the insurance company, the higher the premium.

Explore

Assume you are deciding to insure the contents of an apartment. You must decide on the amount of coverage and the deductible.

The greater the coverage, the greater the premium you pay.
The lower the deductible, the greater the premium you pay.

What probabilities would you like to know to help you decide whether or not to get contents insurance, and if you decide you want it, how much coverage to get and what deductible to choose?

Develop

As well, probabilities based on known facts can be used to help make decisions. Such probabilities are **theoretical probabilities**.

1. Emily is deciding whether to buy a lottery ticket being sold by a charitable organization or to make a donation. To determine the risks and the rewards, she inquires about buying a ticket versus making a donation and she learns the following.

   Each ticket cost $100.
   100 000 tickets are being sold.
   There are 50 valuable prizes to be won.
   Of the $100 ticket price, $25 goes to the charity; the rest goes to pay for prizes.
   Buying a ticket is not making a donation, so no tax receipt is issued.
   Any size of donation would go entirely to the charity and a tax receipt would be issued for the full amount towards reducing her income taxes.

a) The theoretical probability of buying a winning ticket is $\frac{\text{the number of prizes}}{\text{the number of tickets sold}}$. What is the probability, expressed as a decimal, of Emily winning if she buys a ticket?

b) What are the risks and the rewards of buying a ticket?

c) What would you do if you were Emily? Explain.

Practise

2. In the insurance industry, smoking is a risk factor. Smokers pay more for life, homeowner, and contents insurance. What probabilities would be higher for smokers than non-smokers?

3. A store has a semi-annual sale where every customer receives a "Scratch and Save" card offering various discounts.
   One in 1000 cards offers a 50% discount.
   50 in 1000 offer 25%. 200 in 1000 offer 20%.
   749 in 1000 offer 10%.

   a) The theoretical probability of getting a card offering a specific discount is $\frac{\text{the number of cards offering the specific discount}}{1000.}$

   Determine the probability, expressed as a decimal, of each discount.
   i) 50% ii) 25% iii) 20% iv) 10%
   v) a discount of any amount vi) 100%

   b) What are the risks and the rewards for you as a customer in this situation?

For each situation below, answer parts a) to c).
a) What probabilities would you want to know to help make a decision?
b) What are some related risks and rewards?
c) What decision would you make? Why?

4. Neil is tired of wearing glasses. He is thinking about having laser eye surgery to correct his vision.

5. Shana is a wedding planner. She is advising a family on the risks and rewards of holding a wedding outdoors.

6. Daniel just bought a car. He is trying to decide whether to purchase a security system for the car.

7. Sheree is trying to decide whether to invest some of her earnings in Canada Savings Bonds, in a small business that her friends are starting, or in a house.

3.3 Career Focus: Medical Office Receptionist

Margarita is a receptionist in a doctors' office. She enjoys her job which can be very demanding.

Her job involves greeting patients, checking their OHIP cards, getting their files, and taking them to the examining rooms.

She is responsible for answering the telephone—making appointments and answering general inquiries.

Using a computerized scheduler, she schedules appointments for patients to see the doctors in her office, and she arranges for patients to see specialists and to have X-rays and lab work done.

She looks up charts and files for the doctors. As well, she is helping the bookkeeper with the billing.

1. What are the responsibilities of Margarita's job?
2. What skills do you think are needed for Margarita's job?
3. What personal characteristics do you think a receptionist should have?
4. Margarita also directs patients to information in brochures available in the office on topics such as healthy lifestyle and vaccinations. A patient isn't sure about getting a flu shot, or influenza vaccine. Margarita shows him a brochure that points out the following:

   Influenza can be prevented by a flu shot.
   The flu shot protects about 70% of people who receive it.
   Some people get the flu even if they have had the shot, but will be less sick than people who were not vaccinated.
   The vaccine prevents pneumonia in about 6 of 10 elderly people.
   It can prevent death from flu in more than 8 of 10 elderly people.
   There may be mild side effects.

   a) What is the probability of catching the flu if you get a flu shot? Explain.
   b) What benefits does the flu shot offer?
   c) What risks does the flu shot create?
   d) Would you get the flu shot?

5. Another patient is concerned about her child getting the MMR (measles-mumps-rubella) vaccine because she has heard it can cause encephalitis, an inflammation of the brain. Margarita shows her a brochure that points out the following:

One child out of every million children who receive the vaccine gets encephalitis.
Children who do not receive the vaccine could get measles.
If children get measles, they might develop encephalitis.
One in 1000 children who get measles gets encephalitis.

a) What is the probability of developing encephalitis if you receive the vaccine?

b) What is the probability of developing encephalitis if you get measles?

c) What are the given risks and rewards of getting the MMR vaccine?

d) What other probabilities would you like to know before making a decision about getting the MMR vaccine?

3.4 Comparing Probabilities

Explore

The theoretical probability of winning Lotto 6/49 is 0.000 000 071 5. The experimental probability of being struck by lightning is 1 in 500 000. Which is more likely to happen? Some say lotteries are taxes for the uninformed. Explain.

Develop

1. Probability is $\dfrac{\text{the number of favourable outcomes}}{\text{the number of possible outcomes}}$.

You have calculated probabilities using this formula. Identify a favourable outcome and a possible outcome for each probability.

a) The experimental probability of the ice breaking up on particular date(s) is $\dfrac{\text{the number of times on the date(s)}}{\text{the number of years studied}}$.

b) The theoretical probability of getting a card offering a specific discount is $\dfrac{\text{the number of cards offering the specific discount}}{1000.}$

2. A muffin shop has a promotion that indicates you have a one in ten chance of getting a free muffin if you buy a jumbo coffee. On the inside bottom of 5000 of every 50 000 jumbo coffee cups is printed FREE MUFFIN. A favourable outcome is getting a jumbo coffee cup with FREE MUFFIN. A possible outcome is getting a jumbo coffee cup.

a) What is the theoretical probability, expressed as a fraction in lowest terms, of getting a free muffin if you buy a jumbo coffee?

b) Is the muffin shop's promotion correct?

c) What is the probability expressed as a decimal and as a percent?

3. Janine has been waiting to get a free muffin. She has bought 14 jumbo coffees and has not yet received a free muffin. When she bought her 15th coffee, she got a free muffin.

a) What is the experimental probability based on Janine's data?

b) Things are looking up. When Janine buys her 18th coffee, she gets her second free muffin. What is the experimental probability based on 2 free muffins in 18 cups?

c) Do you think if someone was keeping a tally of every person who bought a jumbo coffee and whether or not the person got a free muffin, the experimental probability based on that data would be exactly 0.1 or at least closer to 0.1 than Janine's data? Explain.

Practise

4. **Skills Check** Copy and complete the table.

	Probability			
	Words	**Fraction**	**Decimal**	**Percent**
a)	3 out of 25			
b)	4 in 1000			
c)	2 in 10 000			

5. **a)** Determine the following information about students in your class.
   i) total number
   ii) number of females
   iii) number of each age:
   less than 16   16   17   18   19 or older
   iv) number who wear glasses

   b) Determine the experimental probability that a student chosen at random from your class
   i) wears glasses
   ii) is less than 16 years old
   iii) is less than 19 years old
   iv) is female

   c) Which probability from part b) is greater—ii) or iii)?

6. A hospital lottery advertises a 1 in 8 chance of winning a prize. Eighty thousand tickets will be sold. If the lottery's claim is true, how many prizes is it offering?

7. Canadian Steve Nash is an accurate shooter in basketball. In the 2001–2002 season, the probability of his shots from the floor scoring a basket was about 1 out of 2. His free-throw scoring was 88.3%.

   In the same NBA (National Basketball Association) season, Vince Carter's scoring record for shots from the floor was 43.4%, and his free-throw scoring was about 0.75. Which player had greater accuracy on shots from the floor? on free throws? How do you know?

3.5 Probability Experiments

Explore

The number 7 is considered lucky. If you roll two dice, the probability of rolling a sum of 7 is greater than the probability of rolling any other sum. Show that this is true.

Develop

Some tools are useful for conducting probability experiments.

1. How many possible outcomes are there when you
   a) toss a coin?
   b) toss two coins?
   c) roll a die?
   d) draw a card from a standard deck?
   e) toss a coin and roll a die?

2. **Equally likely outcomes** have the same probability. Which pairs of outcomes are equally likely?
   a) tossing a coin and getting a head; tossing a coin and getting a tail
   b) rolling a die and getting a 6; rolling a die and getting a 4
   c) drawing a card from a standard deck and getting an ace; drawing a card from a standard deck and getting a heart

3. **a)** What is the theoretical probability of getting a 6 when you roll a die?
   b) Roll a die 50 times and keep a tally to determine the experimental probability of getting a 6.
   c) Compare the experimental probability with the theoretical.
   d) Combine your results from part b) with those of your classmates by adding all the times that there was a 6 and adding all the times the die was rolled. Is this experimental probability closer to the theoretical probability than your probability in part b)?
   e) Did you use decimals, fractions, or percents when you compared the probabilities in parts c) and d)? Why?

Practise

4. a) Predict the theoretical probability of tossing three coins and getting the same side up on each, either three heads or three tails.
   b) Toss three coins 50 times and keep a tally to determine the experimental probability of tossing three coins and getting the same side up on each.
   c) Combine your results from part b) with those of your classmates.
   d) You can determine the theoretical probability of tossing three coins and getting the same side up on each by completing a tree diagram like this. Copy and complete the tree diagram. Compare the results with your prediction in part a).

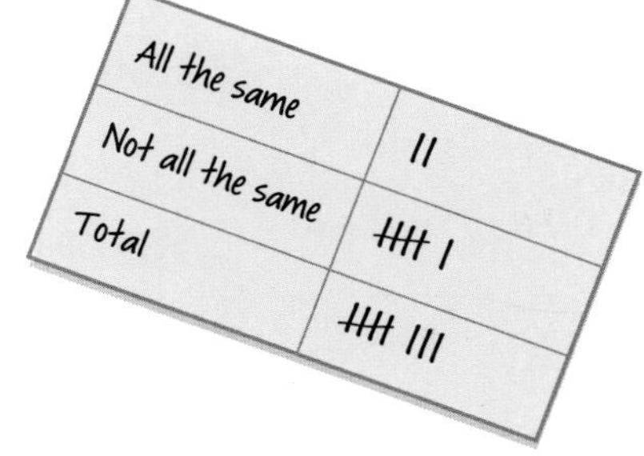

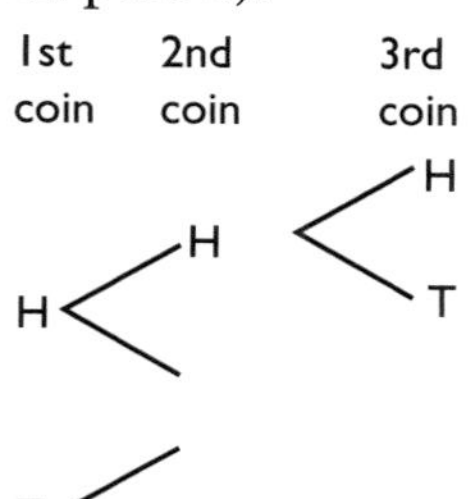

   e) Is the experimental probability in part c) closer to the theoretical probability than your probability in part b)?

5. a) Predict the theoretical probability of rolling two dice and getting two numbers less than 5.
   b) Roll two dice 50 times and keep a tally to determine the experimental probability of rolling two dice and getting two numbers less than 5.
   c) Combine your results from part b) with those of your classmates.
   d) You can determine the theoretical probability of rolling two dice and getting two numbers less than 5 by completing a tree diagram or by completing a chart like this. Copy and complete the chart. Compare the results with your prediction in part a).
   e) Is the experimental probability in part c) closer to the theoretical probability than your probability in part b)?

	⚀	⚁	⚂	⚃	⚄	⚅
⚀	1-1	1-2	1-3	1-4	1-5	1-6
⚁	2-1	2-2	2-3	2-4	2-5	2-6
⚂	3-1					
⚃	4-1					
⚄						
⚅						

3.6 Simulations

Probability experiments can be used to simulate life situations. In a **simulation**, one situation represents another. It often provides a practical alternative to gathering data. For example, if you wanted to determine the probability of a family with four children having all boys, you could locate families with four children and ask how many of the children are boys. Alternatively, you could simulate the situation using a model that provides

- two equally likely outcomes for boy or girl
- groupings of four for the four children

Explore

Use a random number generator or a physical model such as tossing coins or rolling dice to determine the probability that in a family of three children, all three children are the same gender.

Develop

1. A multiple choice test on a topic about which you are completely unfamiliar has ten questions. Each question has four possible answers.

You can determine the probability of passing by guessing, that is, guessing five or more correct answers.

Use a physical model that provides

- four equally likely outcomes for the four possible answers
- groupings of ten for the ten questions

You might use the four aces (or four Kings, etc.) out of a standard deck of playing cards and draw a card ten times.

a) Select the four cards you will use.
b) Decide which of those cards will represent getting a correct answer.
c) Create a tally chart with space for five trials.

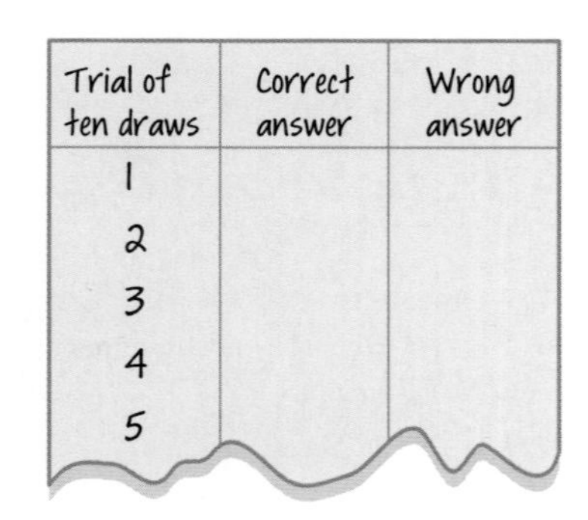

Trial of ten draws	Correct answer	Wrong answer
1		
2		
3		
4		
5		

d) Shuffle the four cards, draw one, record it as a correct or wrong answer, and return it to the others.
e) Repeat part d) for a total of ten draws.
f) Repeat parts d) and e) for a total of five trials.
g) What is the experimental probability of passing the test by guessing?
h) Combine your results from part f) with those of your classmates.
i) Would you believe that the theoretical probability of passing by guessing is 0.078 127? Explain.

2. Rather than drawing all those cards, you can simulate the situation using a random number generator. Because you need four equally likely outcomes for the four possible answers, generate random integers from 1 to 4. Because there are ten questions, generate ten random numbers at a time.
   a) Prepare the random integer generator.
   b) Decide which number will represent getting a correct answer.
   c) Create a tally chart with space for five trials as in question 1.
   d) Record the first seven numbers that are visible as correct or wrong answers. Then, scroll through the numbers until you reach the closing bracket. Record the last three numbers that you could not initially see as correct or wrong answers.
   e) Press enter and repeat part d) for a second trial. Continue until you have five trials.
   f) What experimental probability of passing by guessing did you determine this time?
   g) Combine your results from part e) with those of your classmates. Is this experimental probability closer to the theoretical probability from question 1 i) than your probability in part f)?

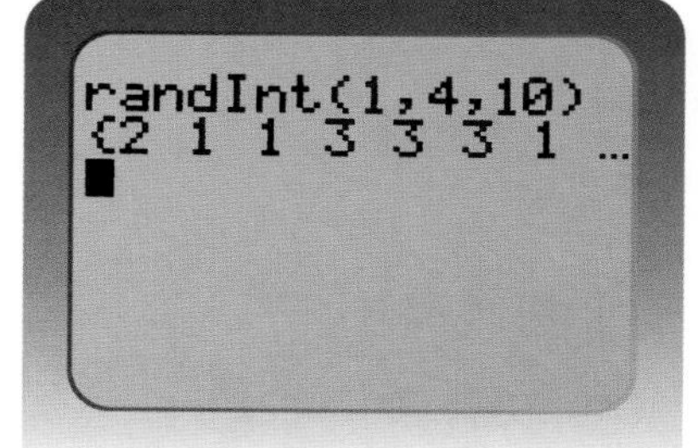

Practise

3. Use a simulation to determine the probability that in a family with four children, all four are boys.
   a) How many equally likely outcomes are there for the gender of a child?
   b) Select a physical model to use.
   c) Decide which outcome will represent a boy.
   d) Create a tally chart with space for ten trials.
   e) Perform ten trials.
   f) What is the experimental probability that in a family with four children, all four are boys?
   g) Combine your results from part e) with those of your classmates.
   h) Would you believe that the theoretical probability of all four children in a family being boys is 0.0625? Explain.

4. Use a random number generator to determine the probability that in a family with four children, all four are boys. Compare your results with those from question 3.

5. **a)** Use a random number generator to determine the probability of winning a Lotto 2/10 lottery, that is, choosing 2 out of 10 numbers correctly. Perform ten trials.
   b) What is the experimental probability of winning a Lotto 2/10 lottery?
   c) Would you believe the theoretical probability is 0.022? Explain.

6. **a)** Use a simulation to determine the probability of answering at least six out of eight questions correctly on a true/false test on a topic about which you are completely unfamiliar. Select a suitable model and perform several trials. (A random number generator is always suitable.)
   b) What is the experimental probability of answering at least six out of eight questions correctly?
   c) Combine your results from part a) with those of your classmates.
   d) Would you believe that the theoretical probability is 0.144 531? Explain.

7. **a)** Use a random number generator to determine the probability that in a family with two children, both children were born in the same month.
   b) What is the experimental probability that in a family with two children, both children were born in the same month?
   c) Combine your results from part a) with those of your classmates.
   d) Why is this model not a perfect match for the situation?

8. Many people believe that if you toss a coin four times and get tails each time, then the next coin is more likely to be heads. Similarly, they believe that if a family has four boys, then the next child born is more likely to be a girl. However, that is not the case. Each toss of the coin or birth of a child is independent of the others. Use a random number generator to disprove this belief.

3.7 Putting It All Together: Spring Birthdays

A What is the theoretical probability that a person selected at random has a spring birthday?

The theoretical probability that three people selected at random all have spring birthdays is only $\frac{1}{64}$, or 0.015 625.

Collect data to determine the experimental probability that three people selected at random all have spring birthdays. Describe what you did. Explain why it is or isn't close to the theoretical probability.

Use a simulation to determine the experimental probability that three people selected at random all have spring birthdays. Describe your simulation. Is it closer to the theoretical probability than the experimental probability you determined above?

3.8 Chapter Review

1. Rissa was at bat 163 times and hit safely 49 times. Taylor was at bat 229 times and hit safely 74 times. For which player is the experimental probability greater for making a safe hit the next time she is at bat? Explain.

2. A charity lottery has 10 000 tickets for $10 each. One first prize has a value of $2000. One second prize has a value of $1000. Five third prizes have a value of $500 each.
   a) What is the theoretical probability, expressed as a decimal, of winning the $2000 prize if you buy one ticket?
   b) What is the theoretical probability, expressed as a decimal, of winning the $1000 prize if you buy one ticket?
   c) What is the theoretical probability, expressed as a decimal, of winning a $500 prize if you buy one ticket?
   d) What are the risks and the rewards of buying a ticket?
   e) Would you buy a ticket? Explain.

3. Cale is trying to decide whether to invest some of his earnings in Guaranteed Investment Certificates, mutual funds, or some blue chip stocks.
   a) What probabilities would you want to know to help make a decision?
   b) What are some related risks and rewards?
   c) What decision would you make? Why?

4. Diane has applied to attend a school that chooses its students at random. She is one of 160 students who have applied for 17 spots. What is the theoretical probability that Diane will be chosen?

5. A competition attracts 7548 people as entrants. Of those, 2173 win a prize. What was the theoretical probability that any entrant would win a prize? Express the probability as a fraction, a decimal, and a percent.

6. In a city-wide promotion, a company advertises that 3% of competitors will win a prize. If 12 694 compete, about how many will win a prize?

7. How many times do you expect heads to appear in 60 tosses of a coin? Explain.

8. **a)** Predict the theoretical probability of tossing three coins and getting only one head.
   b) Toss three coins 50 times and keep a tally to determine the experimental probability of tossing three coins and getting only one head.
   c) Combine your results from part b) with those of your classmates.
   d) You can determine the theoretical probability of tossing three coins and getting only one head by completing a tree diagram like this. Copy and complete the tree diagram. Compare the results with your prediction in part a).

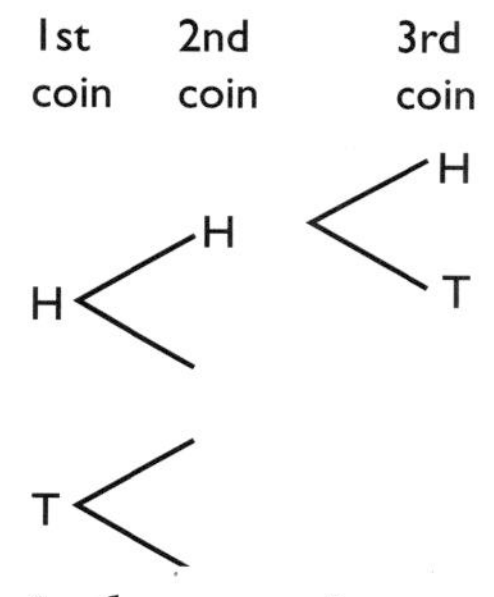

   e) Is the experimental probability in part c) closer to the theoretical probability than your probability in part b)?

9. A brand of cereal enjoyed by children offers one of four different stickers in each box. The chance of getting any one sticker is equal to the chance of getting any other.
   a) Use a simulation to determine each probability.
   - If you buy four boxes, you will get four stickers the same.
   - If you buy four boxes, you will get four different stickers.

   b) What is each experimental probability?
   c) Which probability is greater?
   d) Combine your results from part a) with those of your classmates.
   e) Would you believe that the theoretical probability of getting four stickers the same if you buy four boxes is 0.015 625? Explain.
   f) Would you believe that the theoretical probability of getting four different stickers if you buy four boxes is 0.093 75? Explain.

CHAPTER 4

Renting an Apartment

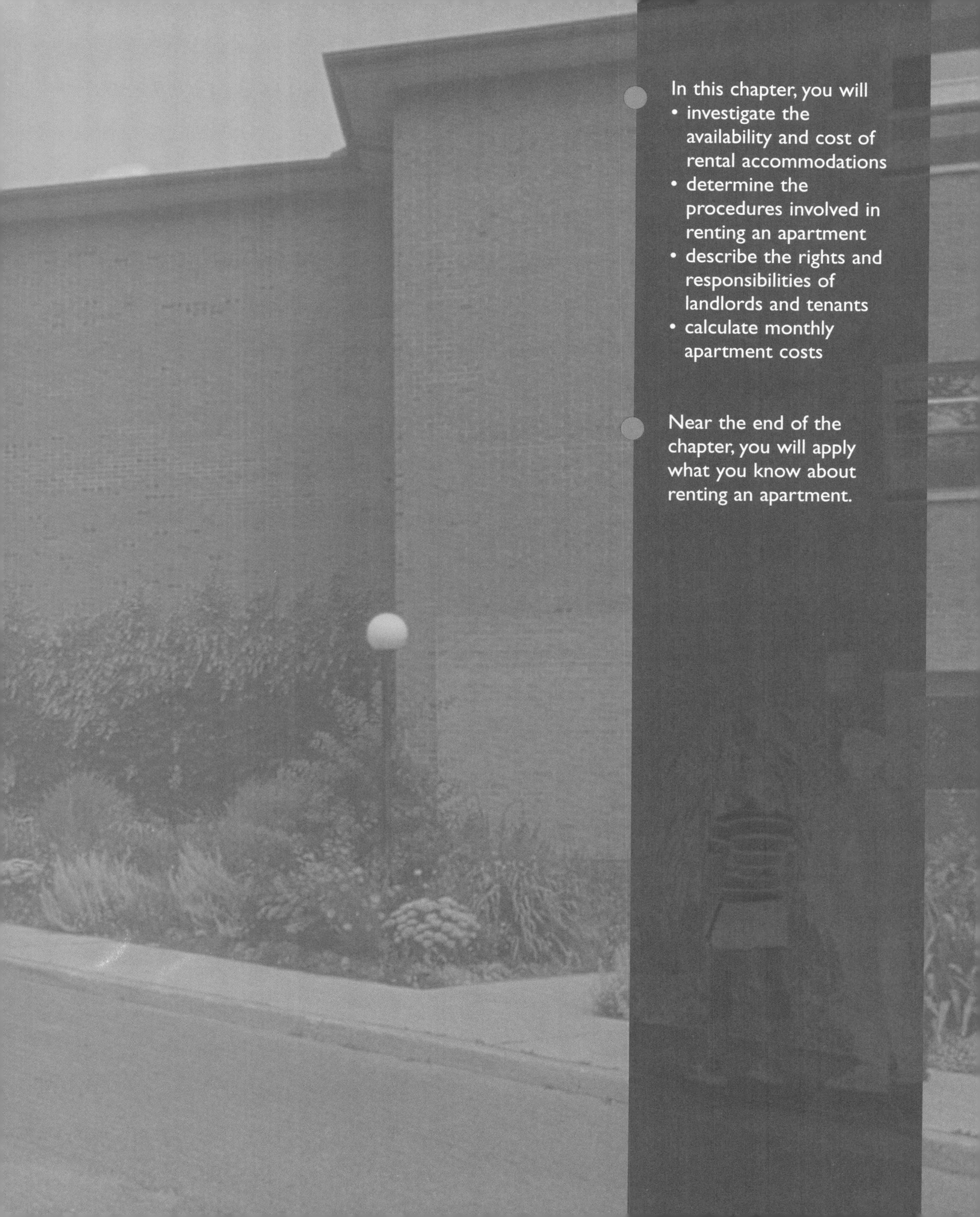

In this chapter, you will

- investigate the availability and cost of rental accommodations
- determine the procedures involved in renting an apartment
- describe the rights and responsibilities of landlords and tenants
- calculate monthly apartment costs

Near the end of the chapter, you will apply what you know about renting an apartment.

4.1 Availability of Apartments

There are many different types of rental accommodations. Rental units exist in townhouse complexes, high-rise buildings, walk-ups, apartments or rooms within homes, shared accommodations, and so on. It is also possible to rent fully detached homes.

Explore

If you were looking for an apartment to rent, what features would be important to you? How much would you expect to pay per month, in your community, to rent the apartment?

Develop

1. Use the Internet, newspaper ads, and/or renters' magazines to investigate types of rental accommodation available in your community. Describe some features you see in the ads.

Access to Web sites for apartment ads can be gained through the *Mathematics for Everyday Life 12* page of www.math.nelson.com.

2. Rental ads show various features of rental accommodations.

a) A high-rise has bachelor, one-bedroom, and two-bedroom units. What is the main difference between a bachelor and a one-bedroom apartment?

b) May thinks she can afford a bachelor apartment for $510 per month plus utilities. The current **tenant**, the person renting the apartment, tells her that utilities totalled about $1000 last year. About how much should May expect to pay for rent and utilities per month?

Fairway/King. Bachelor on ground floor. October. $510/month + utilities. Call 555-8802.

c) A townhouse for rent advertises "4 appliances." What appliances would these likely be?

d) Cable TV is not usually included in the monthly rent. What type of rental accommodation might include cable TV?

e) What other expenses might not be included in the monthly rent?

3. Identify features of apartment rental ads that make the ads effective.

Practise

4. Consider each ad. Identify at least one question you would like to ask. Would you recommend any additional information be given? If so, what?

a) **4-bedroom apartment in downtown area. $1000/month inclusive. Parking extra. 555-1144**

b) **King and University area. $700/month + utilities. Parking included. Available August 1. 555-7357**

c) **Two-bedroom apartment north-east side of city. $800/month inclusive. Parking included. Available September 1.**

d) **One-bedroom. Duke and Charles. Plenty of parking. Available October 1. 555-0783**

e) **Three-bedroom near Bridge and Water. $900/month. Available January 1. 555-3856**

f) **Bachelor. $550/month inclusive. No parking available. Available immediately. 555-1224**

5. Consider your interests and the apartment features below. Classify each feature as
   - required
   - desirable but not required
   - undesirable
   - does not matter

   a) air conditioning
   b) utilities included
   c) pets allowed
   d) dishwasher
   e) underground parking
   f) laundry facilities on site
   g) covered parking
   h) number of bedrooms
   i) outdoor parking
   j) date of availability
   k) outdoor pool
   l) close to shopping
   m) indoor pool
   n) close to park
   o) exercise room
   p) close to walking trails
   q) close to elementary schools
   r) non-smokers only
   s) close to secondary schools
   t) floor level
   u) close to college/university
   v) children allowed
   w) close to public transportation
   x) close to work
   y) other feature
   z) other feature

WS

6. Assume you can afford $800 per month for rent. Use the Internet, newspaper ads, and/or renters' magazines advertising rental accommodation in your community or one nearby.

   a) Identify five ads with apartments that you could rent.
   b) List the amount of the rent for each apartment.
   c) List the features of each.
   d) Identify one question you would like to ask about each.
   e) Select the apartment that you would most like to rent. Give reasons for your choice.

7. Use the Internet or other sources to find two apartments with similar features from different parts of the province. Compare the amounts of the rents. Suggest reasons for the similarity or difference in the amounts.

4.2 Renting an Apartment

Explore

Try this Renter's Quiz.

For questions 1 to 3, select the correct response.

1. How much notice must a landlord give before raising rent?
   a) 30 days **b)** 60 days **c)** 90 days **d)** 6 months **e)** 6 months
2. How much time must have passed since a tenant first moved in or since his or her rent was last increased before the rent may be raised?
   a) 1 month **b)** 3 months **c)** 6 months **d)** 9 months **e)** 12 months
3. How much time do tenants have after signing rental agreements with rent increases to change their minds and tell their landlords they do not agree with the rent increases?
   a) 5 days **b)** 10 days **c)** 15 days **d)** 30 days **e)** 60 days

For questions 4 to 12, answer true or false.

4. For increases in the cost of municipal taxes or utilities, there is no limit on the amount of rent increase that can be charged.
5. Tenants cannot apply to The Ontario Rental Housing Tribunal to have their rents reduced.
6. A landlord can collect fees to process and review rental applications.
7. A landlord is required to give receipts for rent payments, rent deposits, or other charges, if requested by the tenant.
8. The landlord is permitted to charge for receipts given.
9. Mailed rent cheques must be received by the landlord on or before the date that rent is due.
10. Tenants could face eviction if they hold back all or part of their rent because they feel maintenance is inadequate or a necessary repair has not been done.
11. The Ontario Government sets the rent increase guideline every five years.
12. Landlords and tenants of rental units in non-profit, public, or subsidized housing, and college or university residences are not covered by the same rules about rent and rent increases.

Develop

1. Use the Internet to access The Ontario Rental Housing Tribunal Web site or use brochures from The Ontario Rental Housing Tribunal to answer the Renter's Quiz again. Are you surprised by any of the answers?

Access to The Ontario Rental Housing Tribunal Web site can be gained through the *Mathematics for Everyday Life 12* page of www.math.nelson.com.

Practise

2. When a landlord rents to a person, the landlord gives the person a **tenancy**, or a legal right to occupy the rental unit. Every tenant and landlord have a **tenancy agreement**, which may or may not be in writing. A written tenancy agreement is often called a **lease**.

When a tenancy agreement is for a specified period of time with start and end dates set out, the tenancy has a **fixed term**.

a) What is meant by leasing for a year?
b) What would renting month-to-month mean?
c) Rent may be due weekly, biweekly, or monthly. Determine the yearly rent for each apartment described below.
  i) \$825 per month
  ii) \$190 per week
  iii) \$400 biweekly
d) The three apartments in part c) are the same size and have the same features. Why might someone rent the most expensive apartment?

3. Skills Check Calculate each percent, rounded to the nearest cent.

a) 2.5% of \$1100 monthly rent
b) 3.4% of \$800 monthly rent
c) 2.9% of \$750 monthly rent
d) 2.8% of \$675 monthly rent
e) 3.1% of \$1100 monthly rent
f) 3.7% of \$820 monthly rent

4. The Ontario government sets maximum rent increases each year.

Year	Percent Increase
2000	2.6
2001	2.9
2002	3.9
2003	2.9

a) Brendan lived in his apartment for one year before his rental agreement expired on July 31, 2003. His rent was $700/month. What was the maximum monthly rent he could be required to pay for the next year?

b) Barbara lived in her apartment for two years at $850 per month. Her rental agreement expired on February 1, 2002. What was the maximum monthly rent she was required to pay for the next year?

5. A landlord can apply to The Ontario Rental Housing Tribunal to increase rent above the government maximum if the landlord
   - has made major renovations or repairs, or
   - has added security services.

   Rent increases for repairs, renovations, or security services cannot be greater than 4% above the government maximum for rent increase.

   a) Heather's $685 per month rent increased by 2.9% according to the government maximum and by 4% due to added security services. Both increases are calculated on the original monthly rent. What is Heather's new monthly rent?

   b) Chris's rent increased from $700 to $735 per month due to renovations. What percent increase is this? Is this increase within the legal limits?

6. A landlord can collect a deposit from a new tenant. It cannot be more than one month's rent, or, if rent is paid weekly, one week's rent. This deposit must be used as the rent payment for the last month or week of the tenancy. It cannot be used for any other reason, such as paying for cleaning or repairing a rental unit.

   Ron lives in Ingersoll. He finds a job in London and wants to move. He rents an apartment for $800/month. He is required to pay a deposit for the full amount allowed by law.

   a) How much must Ron pay as a deposit?

   b) At the end of his first year of tenancy, Ron's rent increases by 3.9%. What is his new rent?

   c) At the end of the first year of tenancy, the landlord asks Ron to top up his deposit so that it is equal to the new rent. How much should Ron pay the landlord?

   d) A landlord must pay a tenant 6% interest on the deposit every year. How much interest must the landlord pay to Ron at the end of his first year of tenancy?

7. Wilma is relocating to Waterloo. She finds an apartment for \$680/month. The landlord can legally ask for first and last months' rent to be paid by the time Wilma moves in.
   a) What is meant by "first and last months' rent"?
   b) Why would the landlord ask for first and last months' rent?
   c) How much interest on the deposit, at a rate of 6%, must the landlord pay to Wilma at the end of her first year of tenancy?
   d) At the end of Wilma's first year of tenancy, her rent increases by 2.9%. What is her new rent?
   e) At the end of the first year of tenancy, the landlord asks Wilma to top up her deposit so that it is equal to the new rent. How much should she pay the landlord?

8. Omega rents an apartment in North Bay. She has an opportunity to work in Japan for six months. Omega's landlord gives permission for her to **sublet** her apartment. In other words, someone else may live in Omega's apartment until she returns. The lease remains in Omega's name while her friend, Stephanie, sublets the apartment for the six-month period.
   a) Does Omega continue to pay rent to her landlord in North Bay?
   b) To whom does Stephanie pay her rent?
   c) Who is responsible for the apartment?

9. Jack rents an apartment in Sarnia. He is transferred to Niagara Falls when he still has seven months left on his lease. The landlord gives permission for Jack's friend, Trevor, to take over the lease, with Trevor's name on the lease. This arrangement is called an **assignment** of a lease.
   a) Does Jack continue to pay rent to the landlord in Sarnia?
   b) To whom does Trevor pay his rent?
   c) Who is responsible for the apartment?

4.3 Rights and Responsibilities of Landlords and Tenants

In the previous section, you studied rent guidelines and the rights of tenants and landlords with respect to rent. This section will help you become familiar with other issues.

Explore

Try this Renter's Quiz about **Maintenance Responsibilities.**

For all questions, answer true or false.

1. A tenant has to keep the rental unit clean, up to the standard that most people would consider ordinary or normal cleanliness.
2. A tenant is responsible if he or she causes damage to the property, whether deliberate or accidental.
3. A tenant could be evicted for keeping an animal that acted agressively to others.
4. A tenant is not responsible for damage to the property caused by a guest, whether its cause is deliberate or accidental.
5. A tenant is responsible for damage to the property caused by a person subletting the apartment or that person's guests.
6. If something no longer works due to normal wear and tear, the landlord must repair it so that it works properly, or replace it.
7. A tenant can be evicted for holding back rent, even when acting upon a belief that maintenance is inadequate or that a necessary repair has not been done.
8. A tenant can apply to the Tribunal for an abatement of rent—approval to hold back rent—if the landlord has not met expected maintenance or repair duties. The Tribunal order will set out the amount of rent that can be held back and how the tenant should handle this.
9. A landlord must obey all health, safety, and maintenance standards in any provincial laws or municipal bylaws.
10. If rent is late, a landlord can shut off hydro, fuel, or water or interfere with the supply to a tenant.

Try this Renter's Quiz about **Privacy Rights.**

For each situation below, indicate on what terms the landlord can enter. Use the following letter codes.
A: at any time without written notice
B: between 8 a.m. and 8 p.m. without notice
C: between 8 a.m. and 8 p.m. with 24 hours written notice

1. if there is an emergency, like a fire
2. if the rental agreement requires the landlord to clean the unit—unless the agreement allows different hours for cleaning
3. to make repairs or do work in the unit
4. if a notice of termination has been given by either the landlord or tenant, or there is an agreement to terminate the tenancy and the landlord wants to show the unit to a potential new tenant
5. to allow a potential purchaser, insurer, or lender to view the unit
6. if the tenant lets the landlord in
7. to allow an inspection by an engineer or architect or similar professional under the Condominium Act
8. for any reasonable purpose allowed by the rental agreement

Try this Renter's Quiz about **Renewing and Terminating a Lease.**

For all questions, answer true or false.

1. A tenant must move out at the end of a lease.
2. A tenant can stay on past the end of a lease without renewing it, but the rent may go up.
3. If a tenant stays on past the end of a lease without renewing it, all the rules of the former lease still apply to the landlord and the tenant. However, the tenancy is no longer fixed term and the rent may increase.
4. A tenant and landlord can agree to end a lease early without written notice.
5. Daily or weekly tenants must give at least 28 days notice to terminate their tenancy.
6. Monthly tenants must give at least 60 days notice to terminate their tenancy.
7. Monthly tenants do not have to have termination dates at the end of the month.
8. Tenants with a one-year lease may have an earlier termination date than the end of the lease.
9. A landlord may give a tenant a Notice to Terminate if the landlord wants to move into that apartment unit himself.
10. A tenant needs to give only 10 days notice to terminate a lease, if the landlord has given a Notice to Terminate due to upcoming major renovations of the rental unit.

Develop

1. Use the Internet to access The Ontario Rental Housing Tribunal Web site or use brochures from The Ontario Rental Housing Tribunal to answer the Renter's Quizzes again. Are you surprised by any of the answers?

Access to The Ontario Rental Housing Tribunal Web site can be gained through the *Mathematics for Everyday Life 12* page of www.math.nelson.com.

Practise

When a weekly tenant wants to leave, the termination date must be the final day of a weekly rent period. It is recommended that notice be four weekly rent periods as well as at least 28 days.

When a monthly tenant wants to leave, the termination date must be the final day of a monthly rent period. It is recommended that notice be two monthly rent periods as well as at least 60 days.

When a tenant with fixed term tenancy wants to leave, the termination date must be the final day of the agreement.

For questions 2 and 3, select the correct response.

2. Jacek is a weekly tenant with a one-year lease. When the lease expires, he does not renew his lease or sign a new lease. How often will he be expected to pay rent after the expiration date of the lease?

a) weekly **b)** monthly
c) weekly for one year **d)** monthly for one year

3. Zachary is a weekly tenant and pays his rent every Friday to cover rent from that Friday to the following Thursday. On Thursday, March 15, he submits a Notice to Terminate. What is his earliest termination date?

a) Monday, April 19 **b)** Thursday, April 12
c) Monday, May 14 **d)** Thursday, May 17

4. Yolanda is a tenant without a fixed-term agreement. She pays on the first of the month to cover the rent for that month. On July 31, she submits a Notice to Terminate. What is her earliest termination date?

5. William has a year lease that expires on August 31. He does not want to renew his lease and plans to move from his apartment on August 31. What is the latest date for him to give notice?

6. Ute is a monthly tenant. Today her landlord gave her written notice to vacate by September 1, since he intends to move into the unit himself. It is March 15.

Ute knows of a beautiful apartment she can move into immediately, so she decides to vacate as soon as possible. If she submits the Notice to Terminate on March 16, what is her earliest termination date?

7. A landlord can terminate a tenancy only for reasons allowed by the Tenant Protection Act. The landlord must give the tenant a Notice to Terminate a Tenancy.

a) Research and list five reasons that relate to the behaviour of the tenant or the tenant's guests.
b) Research and list five reasons why a tenant may be evicted because of a pet kept by the tenant.
c) Research and list four reasons for eviction that are not based on the behaviour or conduct of a tenant. These are "no fault" reasons.

4.4 Career Focus: Building Superintendent

Jim is the superintendent of a high-rise building. The building is 11 stories high and each floor has 14 apartments. Each floor has 2 small one-bedroom end units, 4 large two-bedroom units, 4 small two-bedroom units, and 4 large one-bedroom units.

Jim is responsible for cleaning all the common areas of the building and maintaining the outdoor pool. He repairs appliances, plumbing, and electrical equipment. He is a troubleshooter, trying to locate and solve small problems before expensive repairs are required. He must work within a budget to hire contractors for major repairs.

1. Suggest a schedule for Jim to follow on a summer day when he will have prospective tenants coming in to view apartments at 11:30 a.m., 3 p.m., and 4 p.m. It is garbage/recycling day. The building's pool is to be open to tenants from 11 a.m. to 9 p.m.
2. Jim is paid $30 000 per year and he has been given an apartment with a market value of $860/month to live in. How much would Jim have to earn a month elsewhere to have the same equivalent "income"?
3. Consider the rent for each size of apartment.
   - small one-bedroom, $650
   - large one-bedroom, $700
   - small two-bedroom, $800
   - large two-bedroom, $860

   Remember that Jim lives in one of the apartments. How much must he collect in rent each month when the building has no vacancies?
4. **Building superintendent required immediately for large high-rise. Previous experience as superintendent an asset. Duties include minor repairs to plumbing and appliances, re-rentals, and complete maintenance of building. Applicant must possess good people skills and enjoy working with seniors. Salary plus apartment.**
   Fax resume to 555-3627.

   What steps could you take in the next few years to become a good candidate for a job like the one in the ad?

4.5 Monthly Apartment Costs

Explore

Most tenants have the following three additional expenses associated with their rented apartments.

- phone
- cable
- insurance

For each of these expenses, investigate the costs of basic coverage, special features, and package deals, if applicable, in your community.

Develop

Most tenants expect to pay monthly telephone and cable TV charges. However, some renters may be unaware that the landlord's insurance covers only the building and not tenants' belongings. Tenants must have their own insurance against losses to their belongings from fire, theft, vandalism, and so on.

1. **Skills Check** Provincial sales tax (PST) applies to some insurance premiums, or what someone pays for insurance coverage. It applies to the type of insurance that renters buy. Calculate the total cost, including PST, on each premium. Round your answers to the nearest cent.

a) $18 per month **b)** $215 per year
c) $196 per year **d)** $178 per year
e) $16.80 per month **f)** $206 per year

2. Natasha estimates the value of the belongings in her one-bedroom apartment at $20 000. Her insurance premium is $162 a year. How much must she budget each month to pay the insurance premium, including tax, on her belongings?

3. Insurance companies suggest that you photograph or videotape your belongings and keep those photographs, along with an inventory list of your possessions, in a safe place (for example, in a safety deposit box). Why would they make that suggestion?

4. Insurance premiums usually cover liability insurance. If someone is hurt in your apartment, you could be held liable, or responsible, and sued. If you are negligent and cause damage to the apartment building or the belongings of another tenant, you could be held liable and sued. Liability insurance protects you from these costs. It is common to have $1 000 000 in liability insurance.
   a) What does it mean to have $1 000 000 in liability insurance?
   b) Insurance policies have a deductible, or the amount you pay for each incident. Which do you think would have a higher premium—a $300 or a $500 deductible? Why?

Practise

5. Cecilia and her husband estimate the value of the belongings in their two-bedroom apartment at $60 000. She has a quote of $299 for insurance coverage for contents and liability for one year.
   a) What is their total bill, including 8% PST?
   b) How much must they budget each month to pay the insurance on their belongings?

6. Aside from $695 per month rent, Halley pays for parking, cable TV, telephone, and contents insurance. On average, he pays
   $30/month for parking
   $27/month for cable TV
   $45/month for telephone
   $35/month for insurance

   a) What are Halley's total monthly expenses for renting and maintaining his apartment?
   b) It is recommended that rent should not be more than the tenant's weekly gross income, or weekly income before any deductions. Halley is paid $1500 biweekly. Is his rent within the recommended amount?

7. Erin and her sister, Moira, share an apartment. They have a category for accommodation in their budget. This category includes all the expenses related to renting and maintaining their apartment.
   a) Use the following information to calculate their monthly accommodation costs.

   utilities (gas, hydro, water), $95
   telephone, $35
   insurance, $30
   laundry, $48
   cable TV, $39
   rent, $725

   b) What combined gross monthly income do they need to be within the recommended amount given in question 6 b)? (Use 4 weeks = 1 month.)

8. André is a security guard. He owns a car and rents an apartment. Below are André's insurance, telephone, and cable TV bills.

Insurance

• Renter's Policy Coverage •	Coverage	Premium
Contents	40 000	$182.00
Personal liability (each occurrence)	1 000 000	10.00
Medical payments to others (each person)	5 000	included
Damage to property of others (each occurrence)	500	included
Credit card/bank card and forgery	5 000	included
Jewellery and furs	2 000	included
Silverware and gold	5 000	included
Home computer	5 000	included
Total annual premium		
PST		
Total		
Policy deductible **$500**		

Telephone

Monthly services.	$41.16
Equipment rentals.	0.00
Chargeable messages.	11.59
Subtotal.	
GST.	
PST.	
Total current charges. . .	

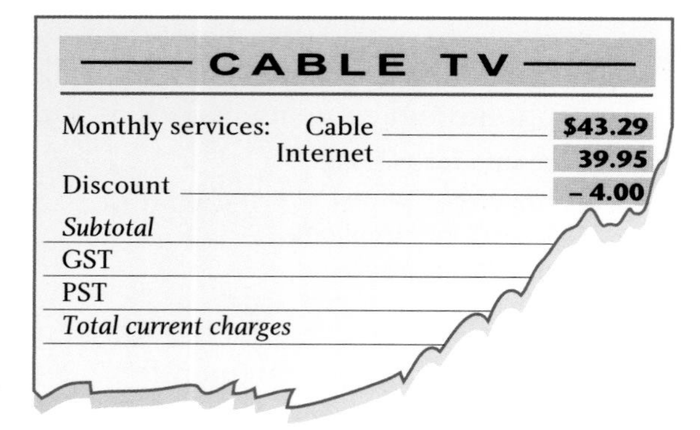

CABLE TV

Monthly services:	Cable	$43.29
	Internet	39.95
Discount		- 4.00
Subtotal		
GST		
PST		
Total current charges		

a) Calculate the taxes and totals for each bill.

b) What are André's monthly premiums for the insurance policy?

c) What are the total charges for insurance, telephone, and cable services for this month?

d) André submits a claim for $2000 worth of damage to his property. About how much money will his insurance company award him?

4.6 Putting It All Together: Renting an Apartment

A You just got a new job in your community and you plan to move from your present living situation. You haven't decided if you will live on your own or share with someone. Your new job pays $24 000 a year. Apply what you have learned in this chapter to complete the following.

- Describe the steps you will follow to find an apartment.
- List features you are looking for and the rent that you can afford both living on your own and living with someone.
- Find a place to rent. Then, indicate which features it has and whether or not you will be on your own.
- Investigate other apartment-related expenses and estimate the cost of each.
- Create a list of your main responsibilities as a tenant.
- Create a list of the main responsibilities of your landlord.

4.7 Chapter Review

1. Consider the following ad for an apartment in a high-rise building.

Large two bedroom. $900/month + utilities. Parking included. Swimming pool and laundry facilities. Convenient location near King and Queen Streets. Available November 1. Call 555-0386.

a) List two questions you would ask the landlord to get more information about the apartment.

b) If utilities cost $85/month, what is the total monthly cost for rent, utilities, and parking?

c) A one-year lease must be signed. By what date must notice be given to terminate tenancy October 31 of the next year?

d) What is the maximum deposit that could be collected?

e) The tenant decides not to terminate the tenancy and renews the lease. The rent goes up by 2.9%. What is the new monthly rent?

f) The rent goes up to $975/month. The apartment underwent some renovations. Does the rental increase stay within the maximum 2.9% yearly increase and maximum 4% renovation increase?

2. List three monthly apartment expenses besides rent.

3. Why is it a good idea to have liability insurance?

4. What are three ways you can find information about apartments for rent in your area?

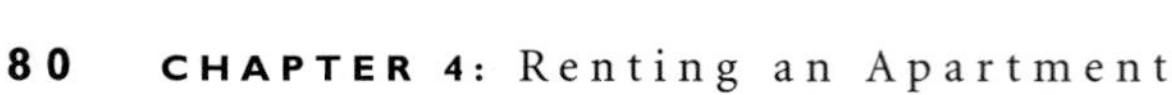

5. **a)** How much of a tenant's income is it recommended that rent should not exceed?

b) How much rent can a person with an income of $32 000 a year pay and stay within the recommended amount?

c) How much would a couple renting an apartment for $1140/month have as their combined annual income if their rent is within the recommended amount?

6. Choose a term to match each explanation below.

walk-up	utilities	bachelor apartment	written notice
lease	rental deposit	sublet an apartment	apartment assignment
eviction	deductible	termination date	contents insurance

Ontario Rental Housing Tribunal

a) new tenant takes over lease with new name on the lease
b) amount by which any insurance claim will be reduced
c) date by which the tenant must vacate
d) apartment without an elevator
e) tenant becomes landlord for new tenant
f) coverage for belongings lost due to theft, vandalism, or fire
g) apartment without bedrooms
h) gas, hydro, and water
i) written tenancy agreement
j) amount up to one month's rent
k) forced to move out of apartment
l) signed form indicating tenant will be vacating the rental unit
m) agency to which tenants and landlords can apply to resolve disputes or gain information about rights and obligations under the Tenant Protection Act

7. Nina places this ad to rent an apartment in her house.

Basement bachelor. $480 incl. utilities & cable. Laundry extra. Street parking by permit.

Nina wants to provide all information related to the cost of living in the apartment. She estimates the following.

rent $480/month	contents insurance $25/month
utilities included	parking permit $30/month
basic cable TV included	basic telephone $30/month
laundry $20/month to use Nina's laundry facilities	

a) Determine the total estimated expenses.

b) What items on the list of apartment-related expenses might vary significantly depending on the tenant?

CHAPTER

5 Buying a Home

In this chapter, you will
- investigate the availability and price of homes in your community
- describe the procedures and costs involved in purchasing a home
- calculate the monthly costs of maintaining a home

Near the end of the chapter, you will describe the procedures and costs involved in purchasing and maintaining a home, given a specific annual income.

5.1 Looking for a Home

Owning a home is something that many Canadians look forward to, and buying a home is the biggest purchasing decision most will ever make.

Explore

Think about your community and nearby communities.

- What types or styles of homes (for example, detached, townhouses) are available?
- What different areas, subdivisions, or neighbourhoods exist?
- What features do people look for in a home?

Develop

Homes can be classified many ways. A basic classification is **new** or **resale**.

Note that **condominium** is a type of property ownership, not a type or style of home. Condominiums can be townhouses, low-rises, or high-rises. Owners of such homes pay a monthly condominium fee towards the cost of insurance and maintaining, repairing, and replacing common elements, such as driveways, yards, roofs, hallways, and lobbies.

WS **1.** Use the Internet, local newspapers, and/or real estate newspapers and magazines to investigate homes for sale in your community and nearby communities.

a) Locate five or more homes for each category. Any one home can be used several times.

- new
- resale
- under $110 000
- between $110 000 and $200 000
- over $200 000
- in one area
- in another area
- in a third area
- detached
- semi or duplex
- townhouse

b) For each category, list the following.

- asking price
- new or resale
- type/style
- address/area
- features advertised

Access to Web sites with homes for sale can be gained through the *Mathematics for Everyday Life 12* page of www.math.nelson.com.

2. Examine your lists from question 1. Are any patterns evident? These questions may help you.
Do new homes have different features than resale homes?
Do homes under $110 000 have different features than homes more than $200 000?
Do detached houses cost more than townhouses?
Do homes in one area cost more than those in another?

3. Describe a typical home that you could buy in your community or a nearby community with each selling price.

a) $130 000 **b)** $200 000 **c)** $280 000

Practise

4. Consider the ads you have seen and what you know about homes.

a) What are advantages of buying a new home and advantages of buying a resale home?

b) What are advantages and disadvantages of condominium ownership?

c) What are the distinguishing features of each type or style of home?

- detached
- semi-detached
- duplex
- townhouse
- mobile

5. "Location, location, location" are the three most important reasons for buying a particular house. What do you think this statement means? Does your work in the **Develop** questions support this statement?

5.2 Buying a Home

Most people, especially first-time homeowners, need to borrow money to pay for a home.

Explore

Try this Home Buyer's Quiz.

For questions 1 to 12, select the correct responses. There is usually more than one.

1. A loan secured by real estate is called a **mortgage**. When you buy a home, you might

a) take out a new mortgage
b) assume an existing mortgage
c) get a vendor, or seller, take back mortgage

2. A common **amortization period**, the time over which the mortgage is repaid, is

a) 15 years **b)** 20 years **c)** 25 years

3. To minimize the amount of interest, or cost of borrowing, that you pay for your mortgage, you should

a) make as large a down payment as you can
b) have a shorter amortization period
c) make payments weekly or biweekly, instead of monthly
d) get a mortgage that allows for lump sum payments

4. A mortgage is repaid in regular payments which are called blended payments. The payments include

a) the principal, or amount borrowed **b)** the interest

5. When you find a home that you want to buy, you

a) sign an Offer to Purchase prepared by your real estate agent, which, if accepted by the vendor, becomes an Agreement of Purchase and Sale
b) pay a deposit of a few thousand dollars

6. The Agreement of Purchase and Sale has a closing date agreeable to the buyer and the vendor. By that time, the buyer's lawyer has arranged the mortgage and searched the title, that is, confirmed that the seller has owned the property being sold. On the closing day,

a) the money goes from the buyer to the vendor
b) the keys to the home go from the vendor to the buyer

7. A guide to how much you can afford to pay for a home is that your monthly **shelter costs** should not be more than 32% of your monthly gross income, or income before deductions. Shelter costs include
   a) mortgage principal **b)** mortgage interest
   c) property taxes **d)** heating costs
8. If a mortgage is for more than 75% of the value of a home, it is called
   a) a conventional mortgage **b)** a high-ratio mortgage
9. Other costs, in addition to interest, associated with obtaining a mortgage are
   a) a mortgage broker's fee if a broker is used to find a lender
   b) mortgage loan insurance premiums and fee; insurance required if the mortgage is for more than 75% of the value of the home
   c) property taxes paid in instalments added to mortgage payments, usually required if the mortgage is for more than 75% of the value of the home
   d) an appraisal fee to have the value of the home assessed; appraisal often required if the mortgage is not insured
   e) a survey fee if the lender requires a survey and you didn't ask the seller for one in the Offer to Purchase
   f) property insurance premiums because the home is the security for the mortgage

10. The taxes that apply to buying a home are
   a) Goods and Services Tax (GST) if a new home
   b) Goods and Services Tax (GST) if a resale home
   c) Land Transfer Tax if a new home
   d) Land Transfer Tax if a resale home
11. Other costs involved in buying a home are
   a) lawyer fees for reviewing the Offer to Purchase/Agreement of Purchase and Sale, drawing up mortgage documents, searching the title of the property, tending to closing details
   b) adjustments to property taxes and utilities if the seller has paid beyond the closing date
   c) moving costs
   d) service hookup fees, for example, phone, electricity
12. Other costs that some home buyers have are
   a) a home inspection fee if they have the home inspected before signing the deal
   b) repairs and renovations, which may be desired or needed whether or not an inspection was done
   c) appliances, furniture, window coverings, other decorating, and tools, for example, lawn mower, step ladder
   d) water quality and quantity certification fee if the home has well service

Develop

The Canada Mortgage and Housing Corporation (CMHC) suggests the following guidelines to estimate an affordable purchase price. The price is based on monthly shelter costs being no more than 32% of monthly gross income. (See Home Buyer's Quiz question 7.)

Annual Gross Income	10% Down Payment	Maximum Purchase Price	25% Down Payment	Maximum Purchase Price
$25 000	$5 400	$54 800	$16 500	$66 200
$30 000	$7 000	$70 000	$21 500	$86 000
$35 000	$8 600	$86 100	$26 500	$105 900
$40 000	$10 200	$102 300	$31 400	$125 800
$45 000	$11 800	$118 400	$36 400	$145 700
$50 000	$13 500	$134 600	$41 400	$165 500
$60 000	$16 700	$166 900	$51 300	$205 000
$70 000	$20 000	$199 200	$61 300	$245 000
$80 000	$23 200	$231 500	$71 200	$284 800
$90 000	$26 400	$263 800	$81 100	$324 500
$100 000	$29 600	$296 200	$91 100	$364 300

1. Sanil and Jasmine together have an annual gross income of $50 000. They want to make a 10% down payment on a home.

a) What is the value of the down payment they would need?
b) What is the maximum amount they should pay for a house?
c) Will they require mortgage loan insurance? Why or why not?

2. **Skills Check** Calculate each percent mentally.

a) 10% of $90 000 **b)** 10% of $125 000 **c)** 10% of $182 000
d) 25% of $90 000 **e)** 25% of $125 000 **f)** 25% of $182 000

3. Land transfer tax varies depending on the purchase price of the property.
For purchase prices up to $250 000,
land transfer tax = 1% of purchase price − $275
Calculate the land transfer tax on a $119 000 home.

Practise

4. Osa has a gross income of $35 000 per year.
   a) What is the maximum he should pay for a house?
   b) He plans to make a 25% down payment. How much must he have for it?
   c) Will he require mortgage loan insurance? Why or why not?

5. List all the costs associated with buying a home as presented in the Home Buyer's Quiz. Identify the costs that all home buyers have. For the other costs, give the conditions under which they exist.

6. Buying a home involves following a variety of procedures that require the help of professionals. Refer to the Home Buyer's Quiz and list all the procedures and corresponding professionals that could be involved.

7. Gino and Maria bought a resale townhouse for $134 000. They paid a deposit of $3000 and at the time of closing paid another $34 000 towards a down payment.
   a) What was their total down payment and how large a mortgage did they require?
   b) How much GST did they pay? Explain.
   c) Did they require mortgage loan insurance? Why or why not?
   d) Calculate the total of their additional costs which were as follows:

land transfer tax	$1065
legal fees	$835
property tax adjustment	$196
appraisal fee	$150
inspection fee	$200
service hookup fees	$86
moving costs	$645

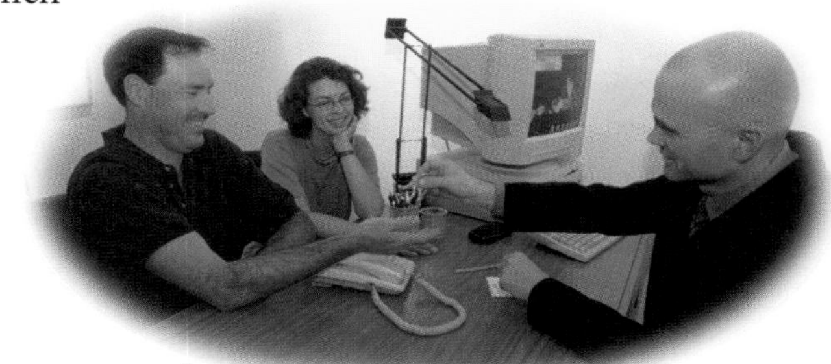

   e) What is the total of additional costs and the down payment?

8. Land transfer tax is calculated using formulas which vary depending on the purchase price of the property.

   For purchase prices up to $250 000,
   land transfer tax = 1% of purchase price − $275

   For purchase prices $250 000 or more and less than $400 000,
   land transfer tax = 1.5% of purchase price − $1525

   For purchase prices $400 000 or more,
   land transfer tax = 2% of purchase price − $3525

   Show why the land transfer tax on Gino and Maria's home from question 6 was $1065.

9. Use the appropriate formula from question 8 to calculate the land transfer tax for a home at each purchase price.

a) $150 000 **b)** $329 900
c) $449 000 **d)** $95 000
e) $51 900 **f)** $399 000

10. Why do you think the land transfer tax formula changes? Do you think this is fair? Explain.

11. Denise and Ryan bought a new home. The price was advertised as $179 900, including GST. What was the price before GST?

12. Sarita and Dante bought a resale home, a condominium unit in a low-rise. It cost $210 000 and they had a down payment of $50 000. Their additional costs were as follows:

legal fees	$735.40
land transfer tax	$
estoppel certificate* fee	$50.00
mortgage loan insurance application fee	$155.35
service hookup of TV cable, telephone, and hydro	$93.22
property insurance premium	$389.89
moving costs	$640.00

* a certificate that outlines a condominium's financial and legal state

a) How large a mortgage did they require?
b) Why did they not pay GST on the purchase price?
c) They obtained mortgage loan insurance. Were they required to obtain it? Explain.
d) Calculate the land transfer tax. (See question 8.)
e) Calculate the total of the additional costs.
f) What is the total of the additional costs and the down payment?

13. Sally and Issac agree to purchase a home for $150 000. What else would you like to know about them and their purchase to estimate the additional costs that they will have?

5.3 The Costs of Maintaining a Home

Explore

You have seen that there are many costs associated with buying a home. Many are one-time costs. Some are ongoing. As well, there are other ongoing costs. List all the costs involved in maintaining a home that homeowners should budget for each month.

Develop

1. A guideline suggests that homeowners set aside about 2% of the purchase price each year for maintenance and repairs. How much should the buyer of a home at each purchase price set aside each year?

a) \$145 000 **b)** \$178 000
c) \$211 000 **d)** \$235 000

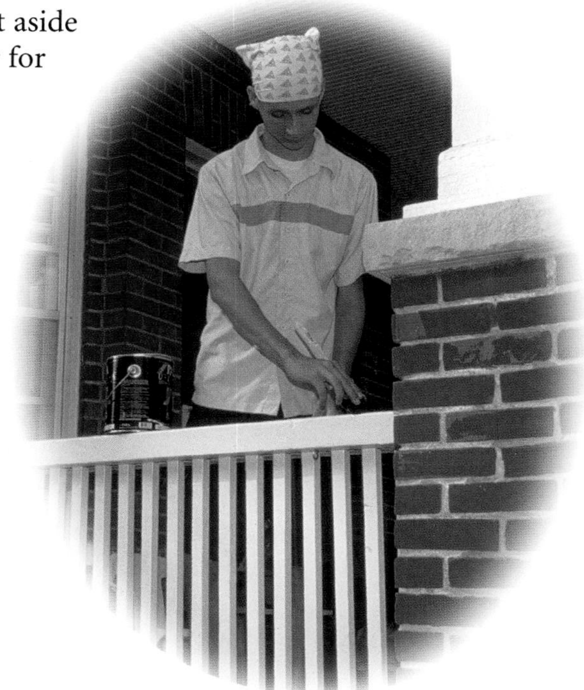

2. a) How would maintenance and repairs differ between new homes and resale homes?

b) You identified the advantages of buying a new home and advantages of buying a resale home in Section 5.1. What might you add to the lists based on what you have learned since?

3. Condominium, you will recall, is a type of property ownership. Condominiums can be townhouses, low-rises, or high-rises. Owners pay a monthly condominium fee towards the cost of insurance and maintaining, repairing, and replacing common elements.

   The fee is based on the size of the unit and pays for
   - some work that other homeowners might do themselves, for example, clearing snow and cutting the grass
   - some utilities such as electricity and water, which other homeowners must budget for as separate items
   - some extras that other homeowners might not have, such as an exercise room, a special function room, or a security system.

   a) How do the costs of maintaining a home that a condominium owner budgets for differ from those budgeted for by other homeowners?
   b) You identified the advantages and disadvantages of condominium ownership in Section 5.1. What might you add to the lists now?

4. The first two costs in Shahnaz's household budget are the mortgage and taxes. Two others are water and snow removal.

   a) Do you think she lives in a condominium? Explain.
   b) List the other costs most likely to be included in her household budget.

Practise

5. Property taxes are paid to the municipality, that is, city, town, or region, where the home is located for services such as garbage and recycling collection, public transportation, parks and recreation, police and fire, libraries, and social services.

   Every property in Ontario has a value attached to it, determined by the Municipal Property Assessment Corporation (MPAC).

   Each municipality sets the tax rate which is a percent of this assessed value.

   Calculate the property taxes, to the nearest cent, on a home with each tax rate and assessed value and determine the amount the homeowner should budget for each month.

   a) 1.0002% of $144 000 **b)** 1.1052% of $158 000
   c) 1.2004% of $203 000 **d)** 1.056% of $258 000

6. a) Under what circumstances is mortgage loan insurance necessary?
   b) Property taxes are usually paid in instalments, added to mortgage payments, under the same circumstances as in part a). What is an advantage of this for the homeowner and for the mortgage lender?

7. Becky and Ron own a home. The costs of maintaining their home are as follows:

mortgage	$773.16 per month
taxes	$2676.86 per year
heating	$930.00 per year
electricity	$125.50 for two months
telephone	$48.20 per month
cable	$51.40 per month
home insurance	$42.24 per month
water	$86.20 for six months
maintenance and repairs	$3600.00 per year

   a) Calculate the amounts per month for the costs not given per month.
   b) Calculate the total monthly costs.
   c) What monthly gross income do Becky and Ron need to be within the 32% guideline with respect to shelter costs?

8. Louise and Jill purchased a new home. The costs of maintaining their home, a condominium unit, are as follows:

mortgage	$914.25 per month
taxes	$3212.48 per year
condominium fee	$312.48 per month
telephone	$48.20 per month
cable	$56.90 per month
home insurance	$42.24 per month
maintenance and repairs	$3600.00 per year

   a) Calculate the amounts per month for the costs not given per month.
   b) Calculate the total monthly costs.
   c) Assume that one third of the condominium fee is for heating. What monthly gross income do Louise and Jill need to be within the 32% guideline with respect to shelter costs?

9. a) Home ownership has many costs that renting does not have. What are they?
   b) List the advantages of owning a home and the advantages of renting.

10. Research the costs of maintaining a particular home. Ask a family member or family friend to share this information with you.

5.4 Career Focus: Real Estate Agent

Alison is a real estate agent. Her job is to bring together people who want to sell a home with those who want to buy a home.

Alison has a variety of skills that have helped her become a successful real estate agent. One is skill in math. She must estimate and calculate to measure and evaluate properties and to help buyers determine what they can afford and what their total costs might be.

Computer skills are also essential as agents, like Alison, list properties, locate properties, and observe trends using computers.

Excellent negotiating skills are also vital. Alison negotiates on behalf of buyers and sellers to arrive at mutually agreeable terms.

If you are selling your home, a real estate agent will
- suggest a selling price for your home by comparing your house to other properties like yours that have sold in your area
- list your property on the multiple listing service (MLS) to reach a large pool of potential buyers
- advertise your home through newspapers and the Internet
- show your home to potential buyers
- negotiate a fair price for your home with the buyer's real estate agent

If you are buying a home, your real estate agent will
- make a list of homes available in your price range and in the area where you would like to buy
- show you homes available for sale
- negotiate a fair price with the seller's real estate agent
- itemize and estimate your closing costs

Alison, like all real estate agents, is paid by commission, which is a percent of the selling price. Since commissions are paid from the selling price, vendors pay real estate agents.

1. The total commission paid is usually 6% of the selling price. This is divided between the real estate agents for the vendor and seller and the companies they work for. A house sells for $152 000. What is the total amount that would be paid in commission?

2. Alison's portion of the commission as the agent for either the vendor or the buyer is usually 1.5%. Her companys also gets 1.5%.
   a) Who gets the other 3%?
   b) Under what circumstances might Alison get 3%?
   c) What is her commission on a home she sold for $152 000?

3. Determine Alison's commission at a rate is 1.5% of the selling price for each selling price.
   a) $160 000 **b)** $240 000
   c) $320 000 **d)** $92 600

4. In June, Alison was the buyer's agent for three homes that she sold for $180 000, $105 000, and $230 000. Her commission rate is 1.5% of each selling price. What was her commission for June?

5. An owner clears $200 000 for his home after the 6% commission is paid. What price did the home sell for?

6. Some vendors try to sell their homes privately, without a real estate agent. Why would they choose to do that? What are the advantages of selling your home using the services of a real estate agent?

7. Research the education and training necessary to become a real estate agent. As well, investigate the advantages and disadvantages of a career in real estate.

Access to Web sites about careers in real estate can be gained through the *Mathematics for Everyday Life 12* page of www.math.nelson.com.

5.5 Putting It All Together: Buying a Home

1. a) Imagine that you and a friend or spouse together have an annual gross income of $60 000 and you have saved $20 000 for a down payment on a home. Determine the purchase price of a home that you could afford.
   b) Use the Internet, local newspapers, and/or real estate newspapers and magazines to find a home that you would like to buy in your price range.
   c) List all the costs involved in buying the home. Estimate the amount of each and the total.
   d) List the monthly costs of owning the home after moving in. Estimate the amount of each cost and find the total.
   e) Recall the guideline that a homeowner's monthly shelter costs should not be greater than 32% of monthly gross income. Do the shelter costs from part d) fall within the 32% guideline?
2. What mathematics do you need to know when it comes to owning a home?

Access to Web sites for housing ads can be gained through the *Mathematics for Everyday Life 12* page of www.math.nelson.com.

5.6 Chapter Review

1. List at least five factors that affect the purchase price of a home.

2. Describe a typical home that you could buy in your community or a nearby community with a selling price of $200 000.

3. Maureen and Ed together have an annual gross income of $70 000. Use the table in Section 5.2 on page 88 to answer the following.

a) Maureen and Ed want to make a 25% down payment on a home. What is the value of the down payment they need?

b) What amount should they be able to afford to pay for a house?

4. What is one cost that buyers of resale homes do not have that new home buyers do?

5. Emma and Joel bought a resale home. It cost $190 000 and they made a down payment of $40 000. Their additional costs were as follows:

legal fees	$685.40
land transfer tax	$
adjustment for taxes paid by the vendor	$356.03
mortgage loan insurance application fee	$145.85
service hookup of TV cable, telephone, and hydro	$86.25
property insurance premium	$325.03
moving costs	$645.00

a) How large a mortgage did they require?

b) They obtained mortgage loan insurance. Were they required to obtain it? Explain.

c) Calculate the land transfer tax. (See Section 5.2 question 8 on page 89.)

d) Calculate the total of the additional costs.

e) What is the total of the additional costs and the down payment?

6. a) List all the costs of maintaining a home.

b) Which three are usually the highest?

c) Which costs are considered shelter costs and used to determine how much buyers can afford to pay for a home?

d) The shelter costs in part c) are sometimes referred to as PITH. Explain.

e) How do the costs of maintaining a home that a condominium owner budgets for differ from those budgeted for by other homeowners?

CHAPTER

6 Household Budgets

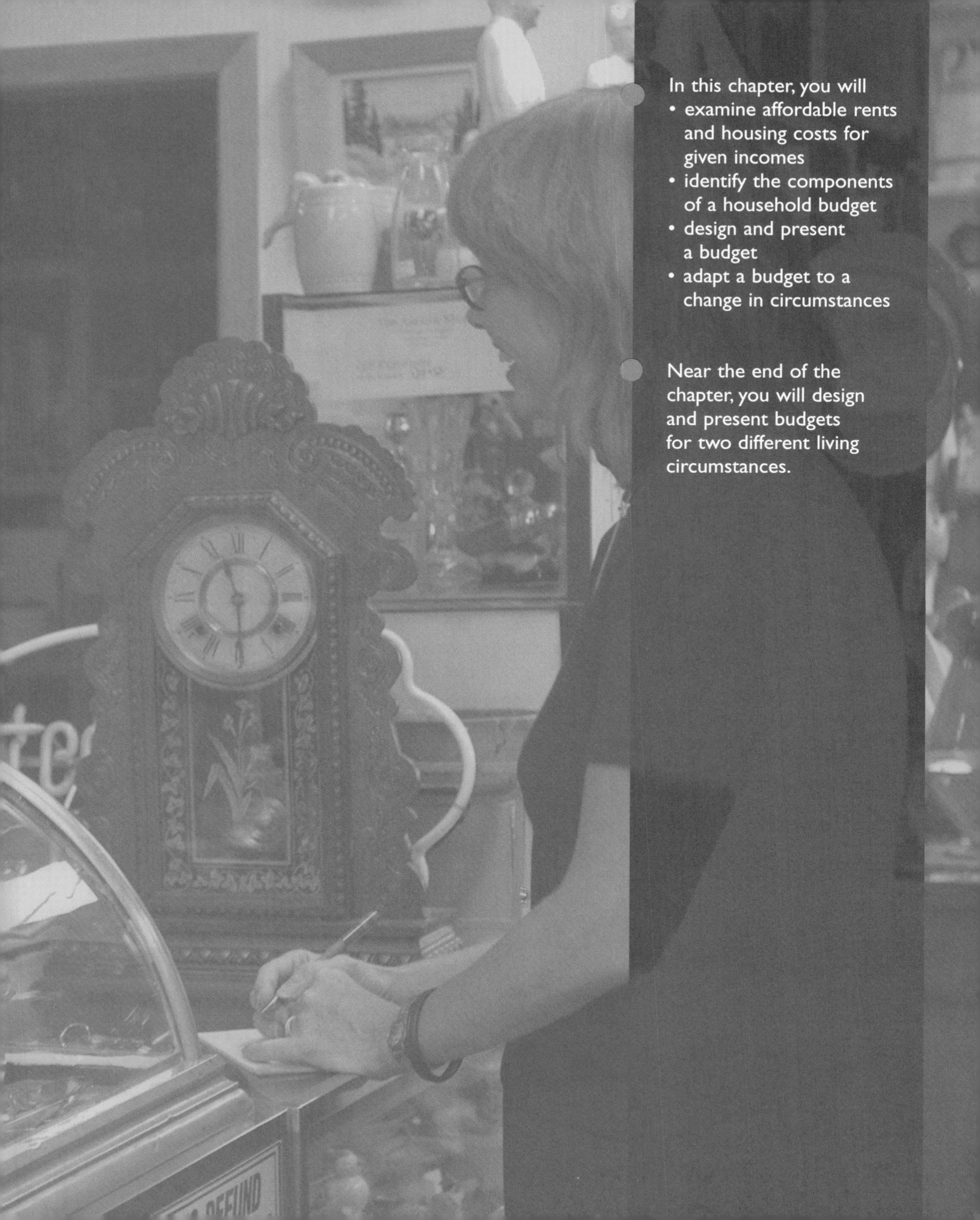

In this chapter, you will

- examine affordable rents and housing costs for given incomes
- identify the components of a household budget
- design and present a budget
- adapt a budget to a change in circumstances

Near the end of the chapter, you will design and present budgets for two different living circumstances.

6.1 Affordable Housing

Explore

Zaib and Aneela are thinking about buying a home.

- They have an annual gross income of $61 000.
- Aneela contributes $28 000 of this.
- Aneela is pregnant and plans to be a full-time caregiver for two years. She expects to receive maternity benefits equal to about two-thirds of her pay for the first year and no pay for the second.
- They currently pay $950 per month for a two-bedroom apartment.
- They have saved $16 000 for a down payment on a home.
- The home they are thinking about has a purchase price of $160 000.

Apply what you have learned in Chapters 4 and 5. What would you advise them to do? Why?

Develop

In Chapter 4, it was recommended that monthly rent should not be more than a tenant's weekly gross income (income before deductions).

In Chapter 5, it was recommended that a homeowner's monthly shelter costs (mortgage principal, mortgage interest, property taxes, and heating costs) should not be more than 32% of the homeowner's monthly gross income. As well, in Chapter 5 on page 88, a Canada Mortgage and Housing Corporation (CMHC) table gave estimates of affordable purchase prices of homes with down payments of 10% and 25% based on shelter costs.

Example 1

Trish and Marshall, who have been married for six years, want to buy a house. Marshall's annual gross income is $27 500 and Trish's is $29 000. They have saved $16 000 for a down payment.

a) According to the table in Chapter 5 on page 88, what purchase price can they afford?

b) How much can they afford for monthly shelter costs?

Solution 

a) annual gross income = Marshall's + Trish's
= 27 500 + 29 000
= 56 500

Their annual gross income and down payment indicate they can afford 10% down.

Annual Gross Income	10% Down Payment	Maximum House Price
$50 000	$13 500	$134 600
$60 000	$16 700	$166 900

They can afford a purchase price of about $150 000 with 10% down.

b) Monthly shelter costs should be no more than 32% of monthly gross income.

monthly gross income = annual gross income ÷ months in a year
= 56 500 ÷ 12
= 4708.33

monthly shelter costs = 32% of $4708.33
= 0.32 × 4708.33
= 1506.67

They can afford about $1500 per month for shelter costs.

CMHC also offers this table which shows the monthly mortgage payment factor for each $1000 of mortgage at various interest rates and amortization periods.

Monthly Mortgage Payment Factors

Rate	Amortization Period			
	25 Years	20 Years	15 Years	10 Years
6.0%	6.398	7.122	8.399	11.065
6.5%	6.698	7.405	8.664	11.311
7.0%	7.004	7.693	8.932	11.559
7.5%	7.316	7.986	9.205	11.810
8.0%	7.632	8.284	9.482	12.064

A monthly mortgage payment can be calculated using the following formula.

$$\text{monthly mortgage payment} = \frac{\text{mortgage}}{1000} \times \text{payment factor}$$

Example 2

a) Determine the monthly mortgage payment for a mortgage of \$135 000 that Trish and Marshall (from Example 1) are considering at 6.5% per year for 25 years.
b) Determine the total amount they will pay and the cost of the mortgage.
c) Suggest ways for them to decrease the cost.

Solution 

a) The monthly mortgage payment factor from the table is 6.698.

$$\text{monthly mortgage payment} = \frac{135\ 000}{1000} \times 6.698$$
$$= 904.23$$

The monthly mortgage payment will be \$904.23.

b)
$$\begin{aligned}\text{total amount paid} &= \text{monthly payment} \times \text{months in a year} \times \\ &\qquad \text{years amortized} \\ &= 904.23 \times 12 \times 25 \\ &= 271\ 269 \\ \text{cost of mortgage} &= \text{total amount paid} - \text{amount of mortgage} \\ &= 271\ 269 - 135\ 000 \\ &= 136\ 269\end{aligned}$$

The total amount paid is \$271 269, and the cost of the mortgage, or the interest, is \$136 269 (more than the mortgage itself).

c) From Chapter 5, mortgage costs can be decreased by
making as large a down payment as possible which they did
having a shorter amortization period
making payments weekly or biweekly, instead of monthly
getting a mortgage that allows for lump sum payments

Practise

1. **Skills Check** Use the given gross incomes to determine the maximum amount that should be spent on rent each month. Where necessary, assume four weeks in a month and a 35-hour workweek.
   a) $1600 per month
   b) $3333 per month
   c) $41 600 per year
   d) $37 440 per year
   e) $8.80 per hour
   f) $14.50 per hour

2. **Skills Check** Use the given gross incomes to determine the maximum amount that should be spent by homeowners on monthly shelter costs. Where necessary, assume four weeks in a month and a 35-hour workweek.
   a) $1600 per month
   b) $3333 per month
   c) $41 600 per year
   d) $37 440 per year
   e) $8.80 per hour
   f) $14.50 per hour

3. Ken rents an apartment in a house for $700 per month.
   a) What is the least annual gross income that Ken should be earning to afford this rent?
   b) What is the minimum hourly rate of pay for a 35-hour workweek that Ken should be receiving to afford this rent?

4. Determine the monthly mortgage payments on a $195 000 mortgage at 6.5% per year amortized over
   a) 10 years **b)** 15 years **c)** 20 years **d)** 25 years

5. Determine the total amount paid on the $195 000 mortgage for each amortization period in question 4. What conclusion can you make? When might you choose 25 years over 10 years?

6. Use the Internet, newspaper ads, and/or renters' magazines to find affordable rental accomodations to meet the needs of each family described below. List any assumptions you make about each situation.
   a) Fiona has just got her first full-time job. Earning $8.65 per hour to start, she will be paid for 35 hours a week for 52 weeks of the year. She has no dependents and does not want to share accommodation.
   b) Eric and Victoria just got married. They both have full-time jobs, with Eric earning $26 000 per year and Victoria earning $28 000 per year. However, they have some debts to pay off.
   c) Alexander is the single father of three-year-old Vienna. He needs child care for Vienna. He earns $25 000 per year.
   d) Nicolas and Christine are the parents of Jack, who uses a wheelchair. Swimming is great physiotherapy for Jack. Christine manages the home; Nicolas earns $50 000 per year working outside the home.

7. Determine whether each mortgage is "affordable" for each gross income at a mortgage rate of 7.5% per year amortized over 15 years.

	Gross Income	Mortgage	Property Taxes and Heating Costs per Month
a)	$48 800 per year	$120 000	$275
b)	$3000 per month	$100 000	$230
c)	$1400 per week	$205 000	$458
d)	$85 000 per year	$203 000	$430

8. Juan and Julie want to buy a home that has more space for their preschool children, Mike and Philipe. Proximity to an elementary school is a priority. Their gross income is $58 000 per year.

Access to Web sites with apartment rental ads and homes for sale can be gained through the *Mathematics for Everyday Life 12* page of www.math.nelson.com.

a) Use the Internet, newspaper ads, and/or real estate newspapers. Find available housing that would be within the affordability guideline for homeowners and would provide the features needed for this family.

b) Choose a down payment. Then, determine the amount of the mortgage.

c) Select an interest rate and amortization period. Then, calculate the monthly mortgage payments based on the interest rate and amortization period selected.

d) Assume property taxes and heating costs will total $375 per month. Show that the mortgage payments plus property taxes and heating costs are less than 32% of the monthly gross income.

6.2 Components of a Household Budget

Explore

Consider your spending that you have been tracking since the start of this course. Categorize how you spend money (for example, entertainment), and list all your spending (for example, movies) in each category.

Which category takes the greatest part of your income? the least?

Which of your spending do you consider necessary? Explain.

Develop

1. a) How do you think your spending will change five years after graduating from high school?
   b) Categorize that spending, and list all the spending that you can imagine in each category.
   c) Which category will take the greatest part of your income? the least?
   d) Which of that spending do you consider necessary?
2. A **budget** is an organized plan for spending money. To make a reasonable plan, you must know your sources of income.
   a) List your current sources of income.
   b) What do you expect your sources of income will be five years after graduating from high school?
3. The items that you spend money on are expenses. Some expenses, called **variable expenses**, change from one month to the next. Other expenses, called **fixed expenses**, are a set amount each month.
   a) Consider all the spending you listed in question 1 b) and identify each as a fixed or variable expense.
   b) How could you estimate how much to budget each month for variable expenses?
4. Some expenses, such as utilities, vary from month to month, and some utility companies offer equal billing. An equal amount is paid each month and any overpayment or underpayment is corrected at the end of each year.
   a) Why would heating bills and electrical bills vary from month to month?
   b) Why would equal billing be useful?
   c) How do you think utility companies decide how much to bill each month?

Practise

5. **Skills Check** Find the average of each of George's variable expense categories.

	Variable Expense Category	March	April	May
a)	food	$70	$85	$73
b)	phone	$43	$52	$39
c)	entertainment	$35	$60	$46
d)	clothing	$55	$67	$280
e)	gifts	$20	$0	$55
f)	miscellaneous	$14	$10	$18

6. **a)** In question 5, which variable expense categories were affected by monthly amounts that were significantly different than those for the other months?
   b) Why might those months have been so different?
   c) How might George use the average amounts when creating his next month's budget?

7. George's fixed monthly expenses are

car payment	$120
car insurance	$85
savings	$135
Internet access	$18

   a) Determine his total fixed monthly expenses.
   b) Given the expenses that George does not have, where might he be living?
   c) Use your results from questions 5 and 6. Calculate the total amount George should budget for next month.

8. Gayle's fixed monthly expenses are

room and board	$225
bus pass	$65
savings	$90
Internet access	$24

Variable Expense Category	June	July	August
phone	$35	$39	$36
entertainment	$85	$60	$49
clothing	$35	$84	$238
gifts	$0	$0	$68
miscellaneous	$18	$15	$22

a) Determine a reasonable amount to budget for each of her variable expense categories.
b) Calculate the total amounts she should budget for next month.

9. Calculate or estimate the amount you have spent on entertainment and one other budget category in the months since you began this course. What amount would you budget for each category for next month? Why?

10. For what months would you need to budget more than average for gifts? Explain.

11. What is impulse buying? How might a budget stop you from buying on impulse?

12. Most financial advisers would recommend that you save 10% of your net income, or income after deductions. Determine the amount that each person should save each week.
a) William's weekly net income is $128 from his part-time job.
b) Carolyn's take-home, or net, pay is $375 per week at her full-time summer job.

13. Think about different living situations, and describe how budgets would be different for each set of situations.
a) rent, own, room and board, live at home
b) live alone, live with someone to share expenses with
c) have children, have no children

6.3 Monthly Budget

Explore

How can you organize all your expenses and income to stay within a budget? What computer software is available to help you do this?

Develop

A budget sheet, budget software, or spreadsheet software can be used to help you plan financially for the month ahead and to record actual income and expenses. These tools enable you to determine if any categories are over or under budget and to determine if your overall spending is within your budget.

1. Ari is a Grade 12 student. He works part time at his parents' restaurant. His monthly net income is $464 plus tips.

Ari's fixed monthly expenses include

savings	$46
car payment	$120
insurance	$85
Internet access	$18

His planned variable expenses include

telephone	$40
gas, oil, and repairs	$40
clothing	$50
entertainment	$50
personal items	$30

Last month, Ari's income was, as he expected, $464, plus $165 in tips. His fixed expenses were as expected. His variable expenses were as follows.

gas and oil change	$70	
clothing	$85	
entertainment	$40	
personal items	$38	
telephone	$35	
birthday presents	$40	
prescription glasses	$55	(amount not covered by family medical insurance)
magazine subscription	$24	
public transit	$5	

Ari used spreadsheet software to record all of his planned, or budgeted, amounts before the beginning of the month. He enters the actual amounts at the end of the month. He calculates the difference between the actual amount and the budgeted amount in each category. A negative sign (−) indicates an amount under budget.

	A	B	C	D
1	**Monthly Net Income**	**Budget ($)**	**Actual ($)**	**Difference ($)**
2	pay from job	464	464	0
3	tips from job	0	165	165
4	**Total**	464	629	165
5	**Expenses**			
6	**Fixed expenses**			
7	savings	46	46	0
8	car payment	120	120	0
9	car insurance	85	85	0
10	Internet access	18	18	0
11	**Variable expenses**			
12	telephone	40	35	−5
13	transportation—bus fare	0	5	5
14	trans—gas and oil change	40	70	30
15	clothing	50	85	35
16	entertainment	50	40	−10
17	personal items	30	38	8
18	medical, dental—glasses	0	55	55
19	other—presents	0	40	40
20	other—subscription	0	24	24
21	**Total monthly expenses**	479	661	182

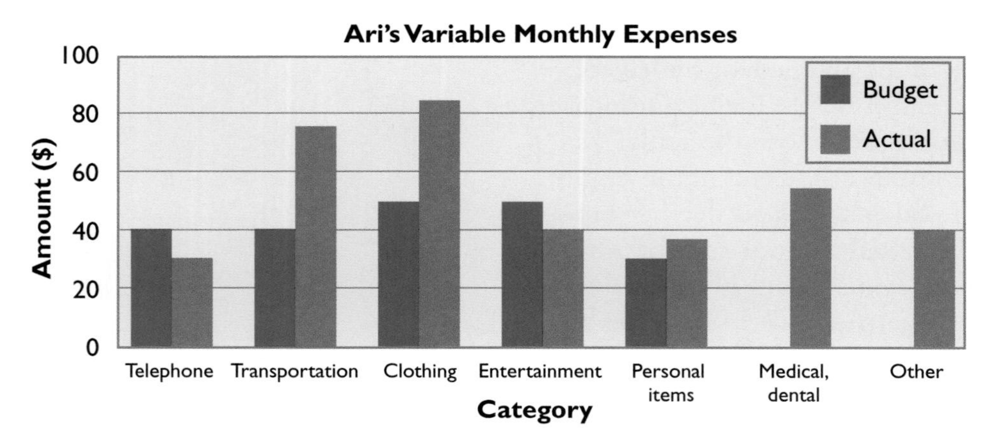

Use the spreadsheet and double bar graph to answer the following questions.

a) How much more are Ari's budgeted expenses than his budgeted income?

b) In what areas did Ari spend more than planned?

c) In what areas did he spend less than planned?

d) Overall, is he over or under budget for the month? Is he still financially all right? Explain.

e) For each of parts a) to d), decide whether it was easier to answer from the spreadsheet or the graph. Why?

Practise

2. Lynne is a high school student living at home. She has a part-time job. She has set up a budget for the month.

a) Calculate her total monthly income.

b) Calculate her total planned expenses.

c) How much more are her planned expenses than her income?

d) Suggest ways she can alter her spending to keep her planned expenses less than or equal to her income.

AT **e)** Complete a balanced budget for Lynne.

Monthly Net Income	Budget
job	$450
investment	$25
Total	

Monthly Expenses	
Fixed expenses	
savings	$45
Internet access	$20
cell phone	$25
transportation—bus fare	$40
Variable expenses	
clothing	$250
entertainment	$50
personal items	$55
school expenses	$35
other—gifts	$50
Total monthly expenses	

3. Vicky has been employed full time for 10 years. She enters her budget in a software program.

Monthly Net Income	**Budget**
job	$3500
investments	$200
Total	$3700

Monthly Expenses	
Fixed expenses	
savings	
mortgage, taxes, and heating	
car payment	$380
car/home insurance	$100
home maintenance	
cable	$40

a) Determine the amount she should budget for savings if she wants to save about 10% of her net income.

b) Determine the amount she should budget for mortgage, taxes, and heating if these shelter costs should be no more than 32% of her $4300 monthly gross income.

c) Determine the monthly amount for house maintenance if she wants to budget 2% of the value of her $135 000 house per year.

d) Vicky created a double bar graph for her budgeted and actual variable expenses, below. For which categories is her spending over budget? under budget?

e) Overall, is her spending over or under budget? How could you check?

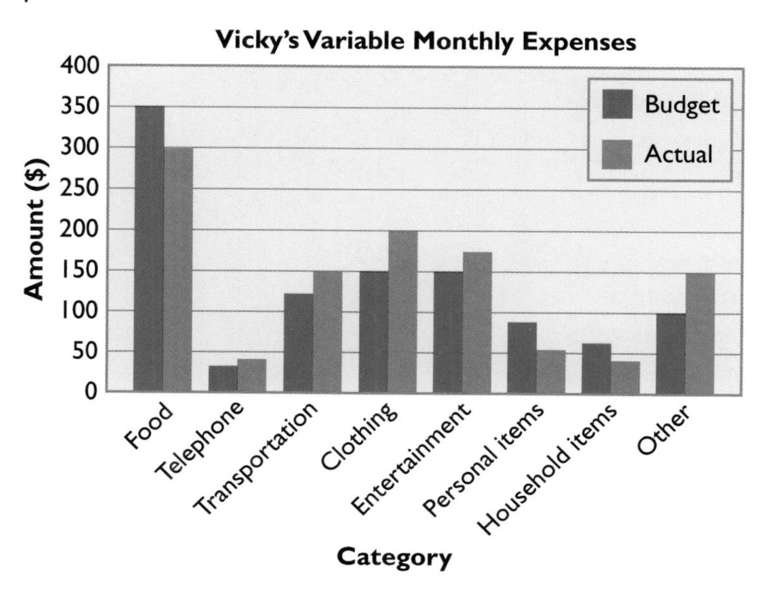

4. **Skills Check** The pie graph shows Monty's expenses for one month.

Monty's Expenses

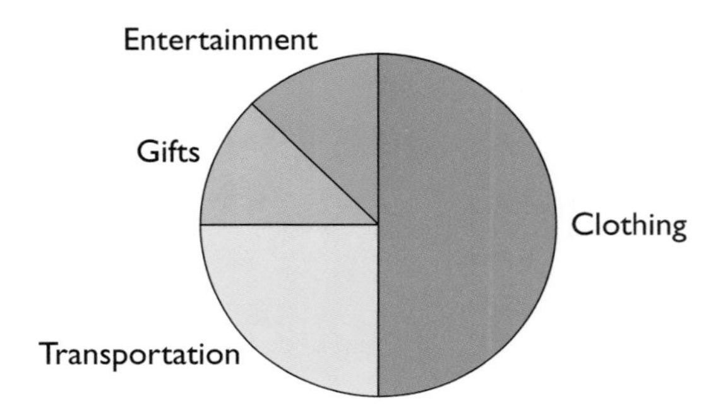

a) In which category did Monty do most of his spending?
b) If Monty spent $25 on entertainment, about how much did he spend on clothing?

5. Matthew has had full-time employment for three years. He has estimated his expenses for the month and entered them on a budget sheet. How much monthly net income must he have to cover his planned expenses?

Monthly Expenses	Budget
Fixed expenses	
savings	$290
rent	$650
parking	$20
car payment	$343
car insurance	$120
cable	$48
Variable expenses	
transportation—gas and oil	$120
food	$360
telephone	$40
clothing	$100
entertainment	$120
personal items	$50
household items	$20
Total monthly expenses	

6. Ajet is a high school student who lives with his family. He has completed all of the entries on his budget sheet except savings.
   a) What is the maximum amount he can budget for savings and keep a balanced budget?
   b) What percent of his monthly net income are his savings from part a)?
   c) If he wanted to save more for a special purpose, what budget items could he adjust?

Monthly Net Income	**Budget**
Total	$640

Monthly Expenses	
Fixed expenses	
savings	
lessons	$160
gym fees	$25
Internet access	$20
Variable expenses	
transportation—bus fare	$40
clothing	$180
entertainment	$80
personal items	$25
school expenses (field trip to Stratford)	$30
Total monthly expenses	

7. Vito is a high school student who lives with his family. His part-time work at a retail store pays $8 per hour. He works a maximum of 60 hours a month. Vito is planning for the upcoming month. His variable expenses will include a concert ticket and a birthday gift for his mother. The expense section of Vito's budget is shown.

Monthly Expenses	
Fixed expenses	
savings	$60
cell phone	$15
Variable expenses	
transportation—bus fare	$40
clothing	$100
entertainment	$150
personal items	$35
other—gift	$65
Total monthly expenses	

a) Vito's net and gross incomes are the same. How much must he make to cover his expenses?
b) How many hours must Vito work in the upcoming month?

AT **8. a)** Prepare a budget for yourself for the next month by completing a budget sheet, or using budget software or a spreadsheet. Revise the budget as necessary until your planned expenses are less than your income.
b) Keep track of your actual expenses throughout the month.
c) At the end of the month, determine whether you are over or under budget.
d) If you are over budget, determine what expenses you could have avoided or put off until a later time.

AT **9.** Dave and Allison, a young married couple, have been working for two years. They have a monthly net income of $3700 and the following monthly expenses.

savings	10% of net income
rent, including utilities	$900
public transportation	$90
clothing	$250
food	$400
entertainment	$100
personal items	$75
prescriptions	$40
household items	$85

Complete the budget column of a budget sheet for them, or use budget software or a spreadsheet to prepare the budget.

6.4 Changing One Item in a Budget

Explore

Matthew, introduced in question 5 on page 112, was in a car accident. Fortunately, no one was injured and his insurance company paid for the car repairs. However, his car insurance premiums increased by $100 each month. Revise his budget so that his total monthly expenses will remain the same. Explain why you made the changes you did.

Develop

It is a good idea to have a budget, but sometimes a significant change in circumstances can throw off even the most careful plan. Then revisions need to be made.

Example

Wendy already had her budget completed when she found out that a $630 car repair bill needed to be included. Revise her budget so that her expenses will still be less than her total net income.

Solution

Monthly Net Income	**Budget**	**Revised Budget**
full-time job	$3200	$3200
part-time job	$500	$500
Total	$3700	$3700

Monthly Expenses		
Fixed expenses		
savings	$520	$520
mortgage payment	$844	$844
car payment	$380	$380
car/home insurance	$200	$200
home maintenance	$320	$320
cable	$40	$40
Variable expenses		
food	$350	$350
utilities	$150	$150
telephone	$35	$35
car—gas, oil, maintenance	$120	**$750**
clothing	$150	**$0**
entertainment	$300	**$40**
personal items	$85	**$5**
household items	$65	$65
other	$100	**$0**
Total monthly expenses	$3659	**$3699**

Practise

AT **1.** Katie has been living with her family and working part time. She uses her parents' car to get to work and public transportation to get to school. After graduation, her employer offers her a raise and full-time employment. She wonders whether she can afford to move out on her own and/or purchase a car. Her budget from last month is shown.

	A	B
1	**Monthly Net Income**	**Budget ($)**
2	**Total**	432
3	**Monthly Expenses**	
4	**Fixed expenses**	
5	savings	50
6	rent	0
7	car payment	0
8	insurance	0
9	home maintenance	0
10	cable	0
11	**Variable expenses**	
12	food—eating out	60
13	utilities	0
14	telephone	0
15	transportation—gas, oil	0
16	trans—maintenance	0
17	trans—bus pass	60
18	clothing	150
19	entertainment	65
20	personal items	35
21	household items	0
22	**Total monthly expenses**	420

With her full-time job, Katie's take-home pay will be $1520 per month. She has the opportunity to share accommodations with a friend at $350 per month (utilities, phone, cable, and parking included). She will pay for her food and her long-distance charges. She sees a car she would like to buy at a cost of $210 per month. She predicts she will have to fill the gas tank once a week, pay $100 per month for car insurance, and budget for maintenance and repairs.

Katie accepts the full-time position. Revise her budget, reflecting the new income. Describe the budget changes and explain whether she can afford to buy the car and/or share accommodations with her friend.

AT **2.** Kyle faces a rent increase of 3.9% in two months. His current monthly budget is shown. Revise his budget to deal with the upcoming rent increase.

	A	B
1	**Monthly Net Income**	**Budget ($)**
2	**Total**	2281
3	**Monthly Expenses**	
4	**Fixed expenses**	
5	savings	290
6	rent	650
7	parking	20
8	car payment	343
9	car insurance	120
10	cable	48
11	**Variable expenses**	
12	transportation—gas, oil	120
13	food	360
14	telephone	40
15	clothing	100
16	entertainment	120
17	personal items	50
18	household items	20
19	**Total monthly expenses**	2281

AT **3.** During a slow time at Kyle's workplace, all employees had their hours cut to 32 from 40. Kyle's monthly net income was reduced to $1824.80 from $2281. Refer to Kyle's budget in question 2. Suggest several items on which Kyle could cut back to ensure that his reduced income covers his expenses, including the increased rent. Kyle does not want to move. Create a revised budget for Kyle.

6.5 Career Focus: Furniture Refinisher

Rita has always been interested in antique furniture and wants to refinish and restore it for a living. Her long-term goal is to own and operate her own company, to be called Rita's Restorations. To achieve her goal, Rita takes woodworking, mathematics, and business English at high school.

Rita works part time for a furniture refinisher while going to high school. She has been introduced to many aspects of the business: the use of brushes and spray guns, various stains, paints, toners, glazes, and sealers, and electrostatic finishing.

Rita prepares furniture for finishing. To strip off old finishes carefully and meticulously, she needs a lot of patience. Her woodworking training gives her the skills to make repairs.

After graduation, Rita plans to work full time for the same furniture refinisher. She will receive on-the-job training and earn an hourly rate of $11.70. After one year, a raise will bring her hourly rate to $12.70.

1. How would Rita's choices of high school courses help her reach her goal of establishing Rita's Restorations?

2. **a)** If Rita is paid for 37.5 hours a week for 52 weeks a year, what is her annual gross income in her first year of full-time work?
b) About how much could she afford to spend monthly on rent?
c) Suppose her net income is 83% of her gross income. What would her monthly net income be?
d) If Rita saved 10% of her net income, how much would she save each month in her first year of full-time employment?

3. What will Rita's annual gross income be in the second year of full-time work? What percent increase is this over the first year's income?

4. Rita has learned that the high average gross income of experienced antique furniture refinishers is about $45 000. If she earned that gross income, about how much could she afford to spend yearly on housing?

6.6 Putting It All Together: Household Budgets

A

AT **1.** Yosuf graduated from high school recently and obtained a job where he earns $12.50 per hour working 40 hours a week. He does not own a car and uses public transportation, costing $45 per month. Assume that net income is about 83% of gross income and that his content and liability insurance premiums will be $210 per year.

Locate affordable accommodations in your community for Yosuf. Explain your choice. Then, create a monthly budget for Yosuf, estimating his fixed expenses and variable expenses where not given. Present the budget using graphs or tables with explanatory notes.

AT **2.** Emma, a construction worker, has a gross income of $36 000 per year. She is a single mother with a two-year-old daughter and pays about $400 per month for child care. She owns a house and pays $600 per year for homeowner's insurance. Her shelter costs are about 28% of her gross income. She owns a car and has $200 per month car payments and $50 per month car insurance payments. Assume that net income is about 83% of gross. One of Emma's goals is to save 10% of her net income.

Create a monthly budget for Emma, estimating her fixed expenses and variable expenses where not given. Present the budget using graphs or tables with explanatory notes.

6.7 Chapter Review

1. Apply the affordability guidelines to the following situations.
   a) What apartment rent could Ketnari consider if her monthly gross income is $2400?
   b) What apartment rent could Mark consider if his yearly gross income is $36 000?
   c) What monthly shelter costs could Ron and Barb afford if Ron's gross income is $45 000 per year and Barb's is $37 000 per year?

2. Quin lives at home with her parents. Her budget for April, May, and June is shown.

Expense Category	April	May	June
entertainment	$45	$18	$120
clothing	$35	$48	$135
gifts	$25	$55	$30
miscellaneous	$15	$18	$21

Determine the average amounts in each expense category. Use these to propose a budget amount for each category for July.

3. a) Give three examples of fixed expenses.
   b) Give three examples of variable expenses.

4. The double bar graph shows Kelly's spending in June.

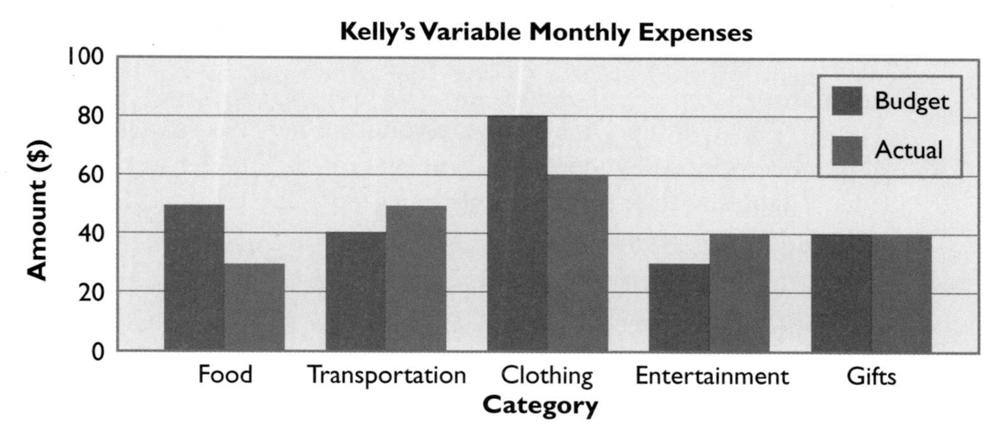

   a) In which categories did Kelly spend over budget?
   b) What is the total amount Kelly had budgeted for the variable expenses shown on the graph?
   c) What is the total amount Kelly spent for the variable expenses shown on the graph?
   d) Overall, has Kelly spent more than her budget in the variable expense categories shown?

5. The budget is Mia's. She is a high school student with a part-time job.

	A	B	C
1	**Net Monthly Income**	**Budget ($)**	**Actual ($)**
2	**Total**	480	480
3	**Monthly Expenses**		
4	**Fixed expenses**		
5	savings	68	40
6	car payment	85	85
7	car insurance	105	105
8	**Variable expenses**		
9	food—eating out	35	45
10	telephone—long-distance charges	20	10
11	transportation—gas/oil	50	48
12	transportation—maintenance	0	30
13	clothing	85	55
14	entertainment		40
15	**Total monthly expenses**		

a) How much can Mia budget for entertainment if she wants a balanced budget?

b) Name the categories labelled i), ii), iii), iv), and v) on the double bar graph to complete the graph showing Mia's budgeted and actual spending amounts in the variable expense categories.

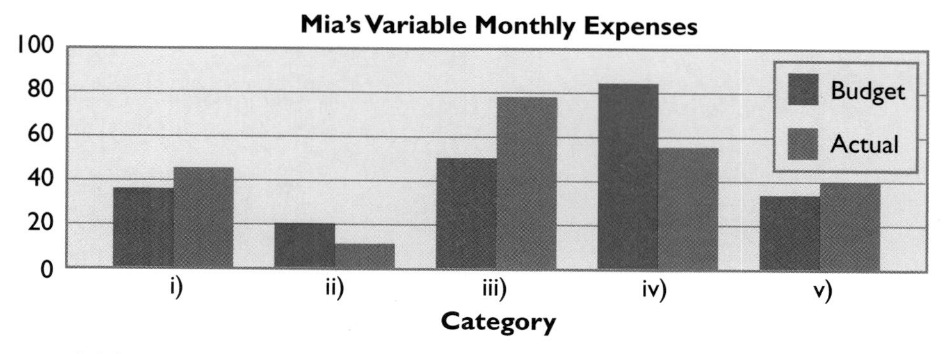

c) In which categories did Mia over spend?

d) In which categories did Mia under spend?

e) Mia tries to save 10% of her income each month. What percent of her income did she actually save?

f) How much money does she have left over at the end of the month?

g) If she put all of that money from part f) into her savings, would she have saved at least 10% of her monthly income?

h) What changes would you suggest she make to her budget? Why?

CHAPTER

7 Measuring and Estimating

In this chapter, you will

- measure lengths and distances in both metric and imperial units
- estimate lengths and distances using personal points of reference
- estimate capacities in metric units
- estimate large numbers illustrated visually

Near the end of the chapter, you will use what you have learned to estimate measurements of buildings in your community.

7.1 The Metric System

Explore

What measurements might you make when
- deciding whether to move to a new place to live?
- moving in?
- living in the new home?

What units of measurement would you use for each measurement you identified?

Develop

To measure an item, it is important to select a suitable unit. Sometimes you may want to express a measurement in more than one unit. For example, Janice measures the height from the floor to the bottom of the cupboard in the home she just bought to make sure she buys a refrigerator that will fit. The height is 175 cm. This can also be expressed as 1.75 m.

You can express a measurement given in larger metric units in smaller metric units by multiplying by the appropriate power of 10: 10, 100, or 1000.

You can express a measurement given in smaller metric units in larger metric units by dividing by an appropriate power of 10.

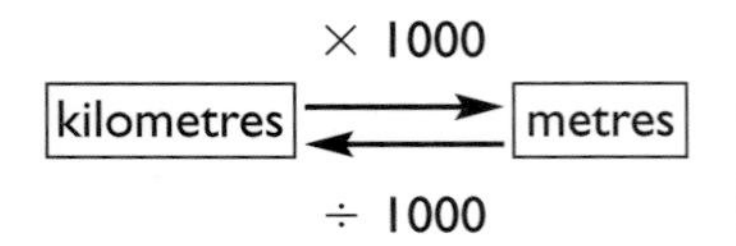

1000 m is equal to 1 km. You can express a measurement given in kilometres as metres by multiplying by 1000.
1.47 km = 1470 m

1. To raise money for a charity, a Grade 12 class walked 6.5 km. What is this distance in metres?

2. What operation would you perform to express a measure given in the first unit as a measure in the second unit?
   a) metres as centimetres
   b) metres as millimetres
   c) kilograms as grams
   d) litres as millilitres

3. Heidi orders 650 g of cheese. What is this mass in kilograms?

kilo
hecto
deka
unit
deci
centi
milli

Practise

4. What metric unit would you use to measure each item?
   a) the length of a skateboard
   b) the length of a room
   c) the length of a ski trail
   d) the thickness of a loonie coin
   e) the mass of an apple
   f) your mass
   g) the mass of a letter
   h) the capacity of a pop can
   i) the capacity of a bathtub
   j) the capacity of a water pail

5. **Skills Check** Multiply mentally.
   a) 38×10
   b) 4.58×100
   c) 2.4×1000
   d) 0.435×10
   e) 4.29×100
   f) 0.38×1000

6. **Skills Check** Divide mentally.
   a) $385 \div 10$
   b) $432 \div 100$
   c) $9436 \div 1000$
   d) $32.5 \div 10$
   e) $43.9 \div 100$
   f) $56.4 \div 1000$

7. Will each number be larger or smaller than the given number? Copy and complete the following.
   a) 3 m = ▢ cm
   b) 28 cm = ▢ mm
   c) 2.4 km = ▢ m
   d) 485 mm = ▢ cm
   e) 4576 m = ▢ km
   f) 35 cm = ▢ m
   g) 2.4 m = ▢ mm
   h) 18 cm = ▢ mm
   i) 2495 mm = ▢ m
   j) 2.4 m = ▢ cm

8. For each pair, which measurement is greater?
   a) 2.7 km or 270 cm
   b) 32.5 cm or 0.435 m
   c) 45 mm or 5 cm
   d) 4007 m or 3.8 km
   e) 4300 mm or 403 cm
   f) 300 km or 30 000 cm
   g) 5 kg or 0.5 g
   h) 0.67 kg or 67 g
   i) 8300 mL or 7 L
   j) 250 mL or 0.025 L

9. a) The length of a lipstick tube is 6.6 cm. What calculation would you do to express this measurement in millimetres?
   b) A lipstick tube measures 2.7 cm in diameter and 6.6 cm in length. If you were making a box to display 24 lipstick tubes in one layer on a store counter, what length and width would you use for the box? Why?
   c) How many of your boxes from part b) could you display on a 0.5 m wide by 1 m long counter?

10. a) The diameter of a quarter is 18 mm. What calculation would you perform to express this measurement in centimetres? Why?
    b) How many quarters could be displayed on a 16 cm by 16 cm plastic sheet if they lie flat side by side?

11. a) Explain how you would express 12 km in metres.
    b) Explain how you would express 535 m in kilometres.

12. a) A basketball player is 196 cm tall. Express this height in metres.
    b) Heidi is 1.53 m tall. What is her height in centimetres?

13. Cornelia's racing car breaks down only 450 m from the finish of a 25 km race. What distance, in kilometres, did she complete before the breakdown?

14. Will each number be larger or smaller than the given number? Copy and complete the following.
    a) 5 L = ▢ mL
    b) 2500 mL = ▢ L
    c) 1.36 L = ▢ mL
    d) 850 mL = ▢ L
    e) 7.1 L = ▢ mL
    f) 5525 mL = ▢ L

15. a) How many millilitres are in 4 L of milk?
    b) How many millilitres are in 1.5 L of water?
    c) What did you do to convert litres to millilitres?

16. a) How many litres are in a 540 mL can of soup?
    b) How many litres are in a 750 mL bottle of ketchup?
    c) What did you do to convert millilitres to litres?

17. Greg is making a tomato sauce for his restaurant. He needs 16 tomatoes, each weighing about 200 g. Tomatoes cost $3.79/kg. How much will the 16 tomatoes cost?

7.2 Measuring Lengths

Explore

What units of measurement are generally used in these stores?

- flooring store
- lumberyard
- clothing store
- fabric store
- paint and wallpaper store
- hardware store

Develop

Before the metric system was introduced in Canada, length was measured in imperial units, such as inches, feet, yards, and miles. These units are still in use. For example, in a football game you might hear that a running back gained five yards on a play, or perhaps a neighbour buys 8 foot two-by-fours at the lumberyard.

The units in the metric system are related by multiples of 10.
The units in the imperial system are related by various multiples.
In the imperial system, 12 inches = 1 foot, or 12" = 1'
3 feet = 1 yard
1760 yards = 1 mile

Example 1

Which imperial unit(s) would be most appropriate to measure each of the following?

a) the height of a basketball player
b) the length of a car
c) the width of a tire rim
d) the distance from Kingston to Hamilton

Solution

a) feet and inches
b) feet or yards
c) inches
d) miles

Smaller units are used for shorter measures, and larger units are used for longer measures. You can thereby avoid the use of either very small fractions or very large numbers of units.

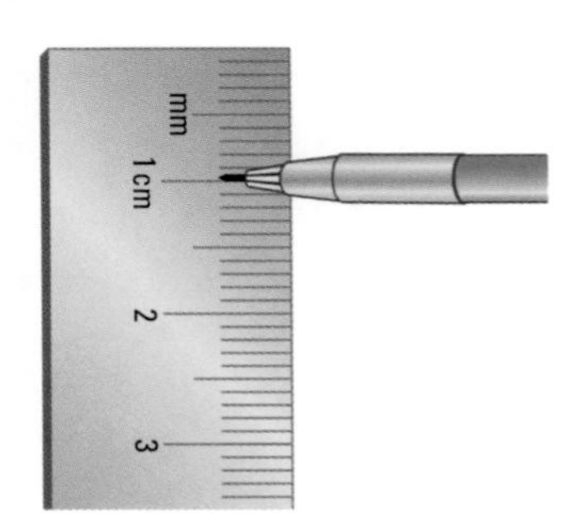

Parts of units in the metric system are expressed as decimals.
For example, the width of a pencil lead is 0.5 mm.

In the imperial system, parts of units are expressed in fractions.
For example, this screw is $1\frac{1}{4}$ inches long.

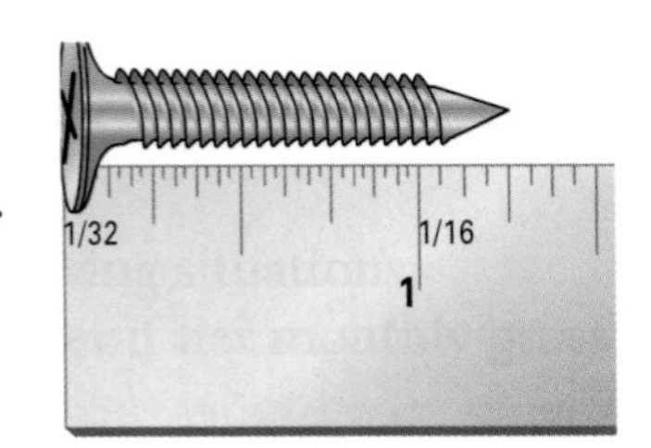

Example 2

What is the length of each nail?

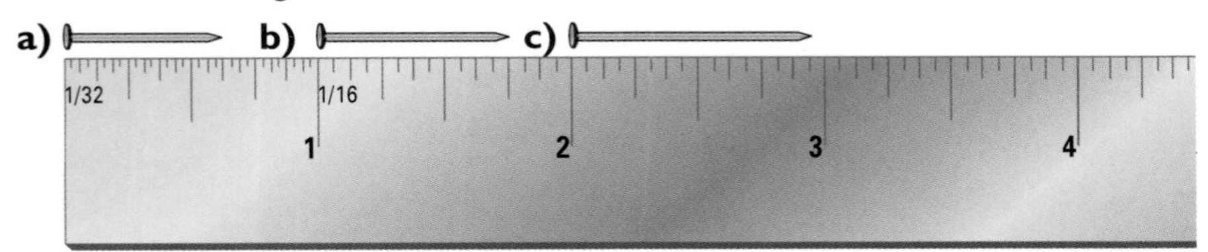

Solution

a) $\frac{5}{8}$ inch **b)** $\frac{3}{4}$inch **c)** $\frac{15}{16}$ inch

Example 3

Measure each distance in inches and in centimetres.
a) the length of an Ontario Health card
b) the diameter of a CD
c) the length of a $5 bill

Solution

a) The length of an Ontario Health card is $3\frac{3}{8}$ inches and it is also 8.6 cm.

b) The diameter of a CD is $14\frac{3}{4}$ inches and it is also 12 cm.

c) The length of a $5 bill is 6 inches and it is also 15.2 cm.

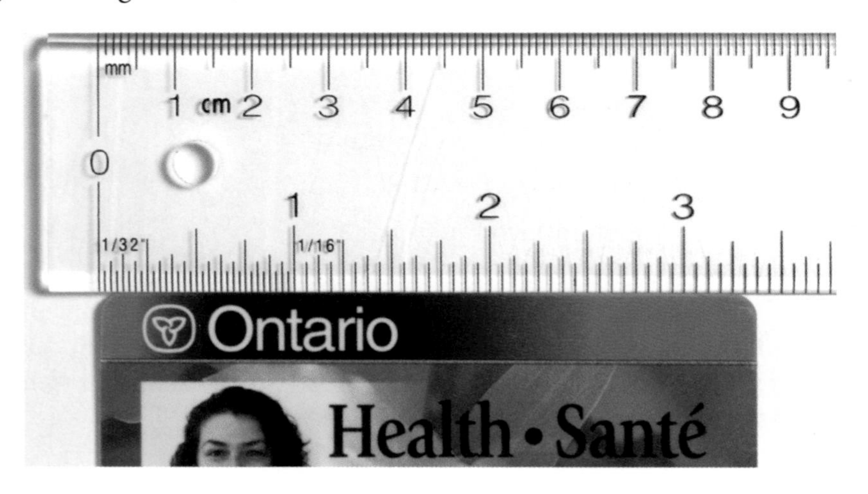

Practise

1. Which imperial unit would be most appropriate to measure each item?
   a) the diameter of a sink drain
   b) the diagonal of a TV screen
   c) the height of a gym
   d) the length of a wallet
   e) the distance from your home to the nearest library

2. Measure each length in inches, feet, or yards—whichever unit is most appropriate.
   a) the width of your thumb
   b) your hand span
   c) the width of your little finger nail
   d) your height
   e) your arm span
   f) the height to your waist from the floor
   g) the height to your shoulder from the floor

3. Measure each length from question 2 in centimetres or metres—whichever is more appropriate.

4. Measure the lengths of the following objects. Express the measurements in both millimetres and centimetres.
   a) a house key
   b) a paper clip
   c) a screw
   d) a marking pen

5. The lumberyard sells wood in imperial units. Plywood is sold in 4' × 4' sheets and 4' × 8' sheets. Explain the dimensions.

6. Cedar deck boards are sold in these "sizes."

   2" × 6" × 8' $\quad$ $\frac{5}{4}$" × 6" × 8'

   2" × 6" × 10' $\quad$ $\frac{5}{4}$" × 6" × 10'

   2" × 6" × 12' $\quad$ $\frac{5}{4}$" × 6" × 12'

   2" × 6" ×16' $\quad$ $\frac{5}{4}$" × 6" × 16'

   a) Explain the dimensions.
   b) How many of which size boards are needed for the floor of a 20 foot by 16 foot deck?

7. **a)** Identify five items sold at a hardware store in imperial units.
   b) Identify five items sold at a hardware store in metric units.

8. Chung's garage measures 5.4 m in length, 3 m in width, and 2 m in height. He wants to buy a van with measurements given in the brochure as 4915 mm in length, 1865 mm in width, and 1710 mm in height. Will the van fit in Chung's garage? How do you know?

9. Raffa has a job making costumes for a dance company. She needs 170 cm of fabric for each skirt required. The fabric costs $12.95/m. Determine the cost, before taxes, of the fabric for nine skirts.

7.3 Estimating Distances

A useful way to estimate the size of an object is to compare it to something whose size you know.

The diameter of a $1 coin is about one inch.

The thickness of a dime is about one millimetre.

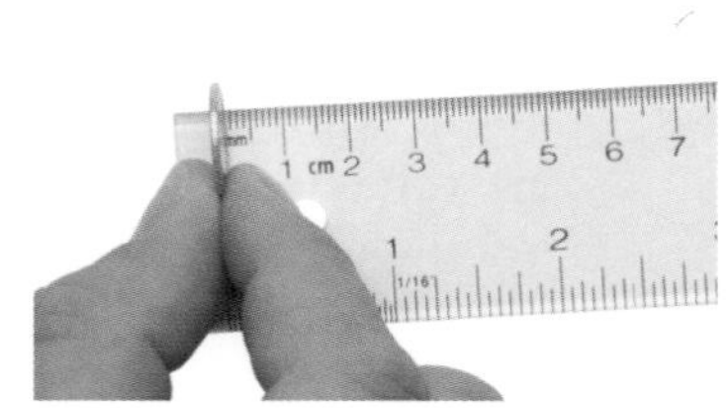

Explore

For each of the following lengths, identify the dimension of an object or a distance that is approximately equal to it.

1 cm	10 cm	30 cm	1 m
1 foot	1 yard	6 feet	

Develop

1. Estimate in metric units. To do so, compare the distance to something you know.

a) the length of a pencil

b) the width of a business envelope

c) the distance around your wrist

d) the length of your arm

e) the length of your classroom

f) the diagonal measure of a computer screen

2. Estimate each measure in question 1 in imperial units. To do so, compare the distance to something you know.

Practise

3. Choose the most suitable estimate for each.
   a) The thickness of a slice of bread is about
      12 mm 12 cm 12 m 12 km
   b) The distance across a street is about
      50 cm 50 mm 50 m 5 m
   c) The height of the CN Tower in Toronto is about
      55 m 550 m 5500 m 55 000 m
   d) The height of a regular doorway is about
      0.2 m 20 m 2 m 0.02 m

4. Estimate in metric units.
   a) the length of a skateboard
   b) the length of a "foot-long" submarine sandwich
   c) the length of a cell phone
   d) the height of a doorknob
   e) the height of a ceiling

5. Choose the most suitable estimate for each.
   a) The width of a party-sized pizza is about
      2 inches 2 feet 2 yards 2 miles
   b) The diameter of a wheel of a 21-speed bike is about
      24 inches 24 feet 2 yards 3 feet
   c) The height of a two-storey house is about
      $2\frac{1}{2}$ feet 25 feet 25 yards 250 inches
   d) The length of a brick used in the construction of a house is about
      8 inches $\frac{1}{2}$ inch 1 foot $\frac{1}{2}$ foot

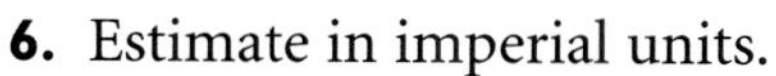

6. Estimate in imperial units.
   a) the height of a door
   b) the height of a binder
   c) the width of a car floor mat
   d) the length of carpet that would fit a school hallway
   e) the length of a van

7. An airline allows carry-on baggage with dimensions up to 55 cm by 40 cm by 23 cm. Find or create a box or object that would fit within the allowable limit.

7.4 Estimating Capacities

Capacity is the amount a container will hold.

As with lengths, a useful way to estimate the capacity of an object is to compare it to something you know.

A 1 L milk carton holds 1 L.

A teaspoon holds about 5 mL.

Explore

For each of the following capacities, identify an object with approximately that capacity.

15 mL　　350 mL　　500 mL　　2 L　　10 L

Develop

1. Choose the most suitable unit—millilitre or litre—to describe the capacity of each. Explain your choice.

a) a pop can
b) a swimming pool
c) a shampoo bottle
d) a washing machine

2. Name at least one type of container that comfortably holds each of the following amounts.

a) 5 mL
b) 250 mL
c) 500 mL
d) 1 L
e) 10 L
f) 50 L

Practise

3. Choose the most suitable measure for the capacity of each item.

a) The capacity of a gas tank in a car is about
5 L　　500 L　　50 L　　5000 L

b) The amount of toothpaste in a new tube is about
0.75 mL　　7.5 mL　　75 mL　　750 mL

c) The capacity of a household aquarium is about
0.7 L　　7 L　　70 L　　700 L

d) The capacity of a soft-drink can is about
0.35 mL　　3.5 mL　　35 mL　　350 mL

e) The capacity of a cereal bowl is about
2 mL 20 mL 200 mL 2000 mL

f) The capacity of a washing machine is about
4 mL 40 mL 4 L 40 L

g) The capacity of a can of paint used to paint a bedroom is about
2 mL 2 L 4 L 6 L

h) The amount of water in a full pail is about
5 mL 50 mL 5 L 50 L

i) The amount of flour used to make a pie crust is about
0.625 mL 6.25 mL 62.5 mL 625 mL

4. Estimate the capacity of each item.
   a) a coffee mug
   b) a kitchen sink
   c) a swimming pool
   d) a standard bathtub
   e) a soup spoon
   f) a four-person hot tub

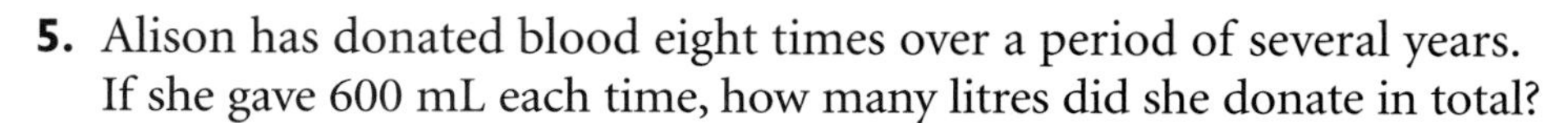

5. Alison has donated blood eight times over a period of several years. If she gave 600 mL each time, how many litres did she donate in total?

6. The instructions on a can of orange juice concentrate indicate to combine the contents with water to make 2 L. If the can contains 500 mL of concentrate, how many litres of water should be added?

7. A health-store owner must fill as many 200 mL bottles as she can with herbal shampoo from a full 40 L container. How many bottles should she be able to fill?

8. In preparing her homemade fruit drink, Sanheel mixes 1.5 L of apple juice, 30 mL of blackberry juice, and 125 mL of prune juice in a 3 L container. How much water must she add to fill the container?

9. What size of paint can would you choose to paint each item? Why?
   a) a picnic table
   b) a living room
   c) a bicycle
   d) a garage floor

7.5 Estimating Large Numbers

Explore

About how many phone numbers do you think are on one page of your phone book? About how many phone numbers do you think are in the whole book?

Develop

Sometimes, you can make a close estimate by estimating with parts. For example, you could estimate the number of people in a large banquet hall by estimating the average number of people at a table and multiplying by the approximate number of tables.

Example

About how many people are in this picture?

Solution

There are about 40 people in one-eighth of the picture. Therefore, there are about 320 people (40×8) in the whole picture.

Practise

1. a) Estimate the number of cars in this mall parking lot. Explain.
   b) About how many people do you think are in the mall? Explain.

2. About how many quarters could fit on this page? How did you decide?

3. a) About how many windows does this 11-floor building have, if opposite sides have the same number? Explain.
   b) About how many people do you think live in the building? Explain.

4. Explain how you could estimate the number of books on the shelves
   a) in your classroom
   b) in the school library

   Estimate each number of books.

5. Identify any assumptions you made in questions 1 to 4. How do the assumptions affect the accuracy of your estimates?

7.6 Career Focus: Decorating Store Clerk

Corrine works at a store that sells window coverings. She loves her work helping people select the style, colour, and fabric of window coverings.

All window covering are custom made and precise measurements are essential. Once customers have decided on the style, Corrine instructs them where to measure their windows.

For inside-mount blinds, the glass itself, not any frame, is to be measured. To determine dimensions to order, Corrine must subtract $\frac{1}{8}$ inch from both the length and width measured by the customer. Then she looks up the sizes in the order book for the chosen style to determine the cost to the customer.

1. A customer orders the following inside-mount blinds style 3402, colour 36–taupe.

3 blinds at 52 inches $\times$ $39\frac{7}{8}$ inches

1 blind at 80 inches $\times$ $70\frac{5}{8}$ inches

3 blinds at $56\frac{1}{4}$ inches $\times$ $42\frac{3}{4}$ inches

1 blind at $52\frac{3}{4}$ inches $\times$ $36\frac{5}{16}$ inches

a) Determine the measurements (length $\times$ width) of the blinds that Corrine will order. Use a ruler to help you.

b) Determine the price of each blind to be ordered.

Style 3402 Prices		**Width** (inches) greater or equal to first number, less than second number			
		10–30	30–50	50–70	70–90
Length (inches) greater or equal to first number, less than second number	10–30	$42	$84	$126	$168
	30–50	$78	$162	$186	$252
	50–70	$120	$180	$236	$276
	70–90	$162	$248	$270	$312

c) Both PST and GST apply. Calculate the total cost of the 8 blind order.

d) Corrine requires a 40% deposit on the order. How much must the customer pay now? at the time of pickup?

7.7 Putting It All Together: Estimating

1. Choose a brick building in your community.
   a) Estimate the number of bricks on the outside walls of the building.
   b) Measure the length and height of one brick.
   c) Estimate the dimensions of the building, based on the dimensions of the brick.
   d) Did you use metric or imperial units? Explain your choice.

2. Choose a building in your community that is not brick.
   a) Estimate the dimensions of the building.
   b) Estimate the number of bricks that would be needed to cover the outside of the building.

7.8 Chapter Review

1. Give the most suitable metric unit of measure to use for each of the following.
a) the width of a hand
b) the width of a hall in your school
c) the distance run by a cross-country runner
d) the height of a person
e) the diameter of a drinking straw

2. Copy and complete the following.
a) 2 m = ☐ cm
b) 48 cm = ☐ mm
c) 3.2 km = ☐ m
d) 392 mm = ☐ cm
e) 3450 m = ☐ km
f) 48 cm = ☐ m

3. Measure the diameter of each coin using suitable imperial and metric units.
a) penny
b) nickel
c) dime
d) quarter
e) $1 coin
f) $2 coin

4. Measure each distance using suitable imperial and metric units.
a) the length of this book
b) the width of your classroom
c) the height of the chalkboard above the floor

5. Bernie decides to start jogging each day and jogs 3 km on the first day. He increases the distances he jogs by 600 m each day.
a) How far will he jog on the fourth day?
b) How far does he jog altogether after 4 days?

6. Choose the most suitable measure for each.
a) the length of a toothbrush
16 m 16 cm 16 mm
b) the length of a paper clip
3 mm 3 cm 3 m
c) the length of a newborn baby
50 cm 5 m 50 mm
d) the height of a house
8 m 0.8 m 80 m

7. What is the most suitable imperial unit for each measure?
a) the width of a CD
b) the height of a person
c) the thickness of a magazine
d) the height of the ceiling in a house
e) the distance travelled by a car in one day

8. Choose the most suitable measure for each.

a) the width of a window

3 inches $\quad 36\frac{1}{2}$ inches $\quad$ 360 inches

b) the length of a mattress

3 yards $\quad$ 37 inches $\quad$ 7 feet

c) the height of a telephone pole

9 inches $\quad$ 9 feet $\quad$ 9 yards

9. Estimate each dimension in both imperial and metric units.

a) the length of a car

b) the width of a car

c) the height of a ceiling in a school

10. Choose the most suitable measure for each.

a) the amount of gasoline in a full motorcycle tank

13 mL $\quad$ 13 L $\quad$ 130 L

b) the amount of freshly poured coffee in a mug

8 L $\quad$ 10 mL $\quad$ 250 mL

c) the amount of dishwater in a kitchen sink

15 mL $\quad$ 15 L $\quad$ 1.5 L

11. Copy and complete the following.

a) 3 L = ■ mL $\quad$ **b)** 1.25 L = ■ mL $\quad$ **c)** 750 mL = ■ L

12. About how many stitches would be in this needlepoint picture, whose actual dimensions are 21 cm by 16 cm. How did you make your estimate?

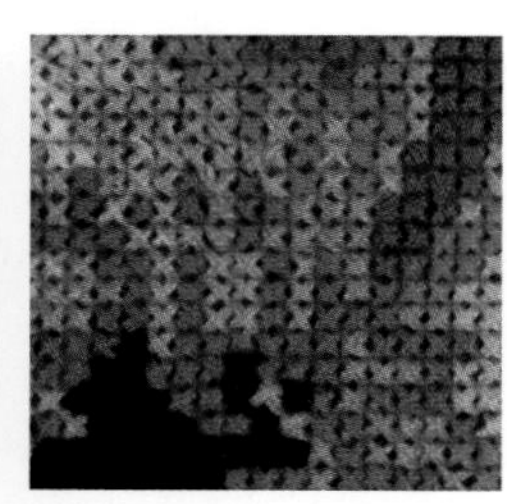

Actual size

13. a) Why is it important to know how to use both the imperial and metric systems of measurement?

b) Most countries use the metric system. Why do you think that is?

CHAPTER

8 Measurement and 2-D Design

In this chapter, you will

- apply the Pythagorean theorem to construct regions with square corners
- calculate areas and perimeters from diagrams
- estimate areas and perimeters of large regions
- investigate the effect of changing the dimensions of shapes on their areas
- make two-dimensional scale drawings

Near the end of the chapter, you will apply what you have learned to design a playground.

8.1 The Pythagorean Theorem

Explore

Use two pieces of string—one 3 feet long and the other 4 feet long.

Lay the pieces on the floor. Pull them out straight. Hold one end of each piece together. Move one piece until the angle created where the ends meet is a right angle. How can you check that it is a right angle?

Measure the distance between the free end of one piece and the free end of the other.

Develop

1. The 3-4-5 triangle is a special triangle. Draw or use geometry software to do the following.

a) Draw a line segment 3 cm long near the middle of a sheet of paper.

b) At a right angle to one end of the 3 cm line segment, draw a line segment 4 cm long. How can you be sure you are drawing the line segment at a right angle?

c) Complete the triangle by drawing the other side. What length do you think it will be? Measure it to check.

d) Draw a square on each side of the triangle. Make sure the corners meet at right angles.

e) Use $A = s^2$ (Area = side × side) to calculate the area of the three squares.

f) Identify the relationship among the areas of the three squares.

The **hypotenuse** is the longest side of a right triangle. It is the side opposite the right angle.

The **Pythagorean theorem** explains the relationship among the lengths of the sides of a right triangle: The square on the hypotenuse is equal to the sum of the squares on the other two sides.

$$\text{hypotenuse}^2 = (\text{side } a)^2 + (\text{side } b)^2$$

$$h^2 = a^2 + b^2$$

a h b

Example

The sides of a rectangular field are 36 m and 21 m, as shown. Meg wants to walk from P to R. Find the distance, to the nearest tenth of a metre.

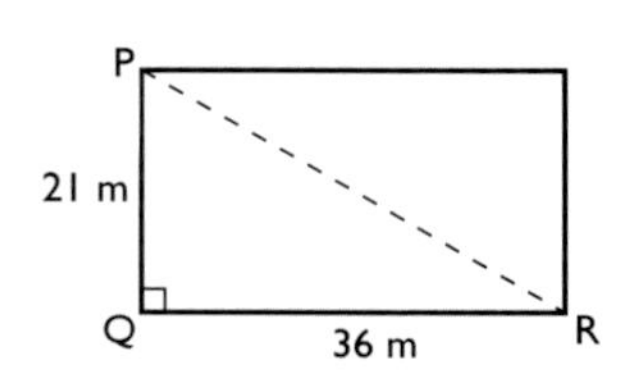

Solution

The distance from P to R is the hypotenuse of the right triangle PQR.

$$\begin{aligned} h^2 &= a^2 + b^2 \\ &= 36^2 + 21^2 \\ &= 1296 + 441 \\ &= 1737 \\ h &= \sqrt{1737} \\ h &\doteq 41.7 \end{aligned}$$

The distance from P to R is 41.7 m.

Practise

2. Skills Check Evaluate.

a) $12^2 + 18^2$ **b)** $14^2 + 9^2$ **c)** $22^2 + 18^2$
d) $4.5^2 + 7.8^2$ **e)** $23.5^2 + 11.2^2$ **f)** $13.6^2 + 3.0^2$

3. Skills Check Evaluate. Round your answers to one decimal place.

a) $\sqrt{45}$ **b)** $\sqrt{80}$ **c)** $\sqrt{120}$ **d)** $\sqrt{200}$ **e)** $\sqrt{250}$

4. Calculate the length of the hypotenuse of each triangle. Round your answers to the nearest tenth of a metre, if necessary.

a)
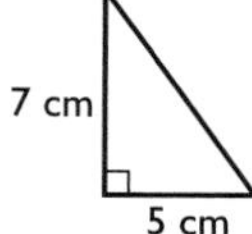

b)
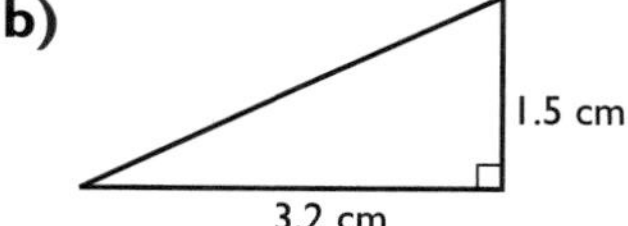

c)
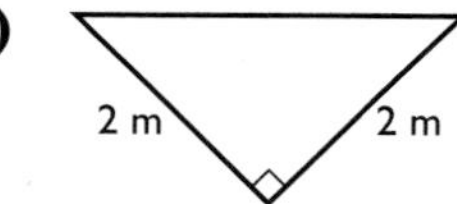

5. Jasna is making a kite with the dimensions shown. What length of wood, rounded to the nearest tenth of a centimetre, is needed for the support indicated?

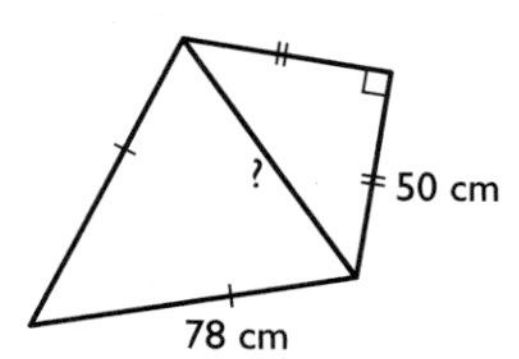

6. Felix is building a garage on a floor that measures 18 feet by 24 feet.

a) Calculate the length of the diagonal of the rectangular floor.

b) Felix measures the length of the diagonal to be $29\frac{1}{2}$ feet. Are the angles at the corners of the garage right angles? Explain.

7. Velta is laying out forms for the cement footings of a house. The house is to be 36 feet by 48 feet. When Velta measures the length of the diagonal of the rectangle, how long should it be so that there will be right angles at the corners of the house?

8. Angelo is going to mark off a rectangular playing field measuring 5 m by 11 m. Explain how he could use the Pythagorean theorem to make sure that the angles at the corners of the field are right angles.

8.2 Calculating Perimeter and Area

Explore

Larissa and Kerry are going to redecorate their L-shaped living/dining room. They want to lay pre-finished hardwood on the floors and install new baseboards all around. Help them to determine the amount of both materials that are required.

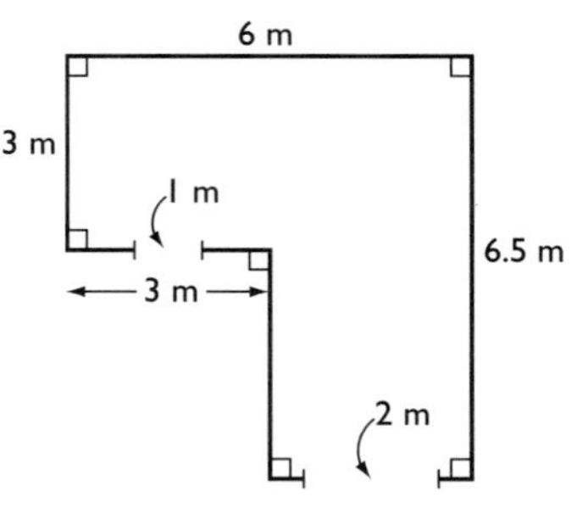

Develop

Example 1

A fence is being installed around this rectangular field. How many metres of fencing are required? Why might more than that amount be ordered?

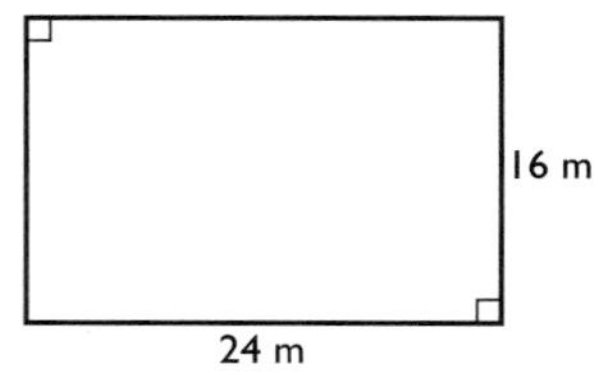

Solution

Use the perimeter of a rectangle formula.

$$\begin{aligned} P &= 2l + 2w \\ &= 2(24) + 2(16) \\ &= 48 + 32 \\ &= 80 \end{aligned}$$

80 m of fencing are required. More than that might be ordered to allow for waste and overlap, or because the fencing may only be available in certain lengths.

Example 2

This lunch room in an office building is going to get ceramic tiles installed.

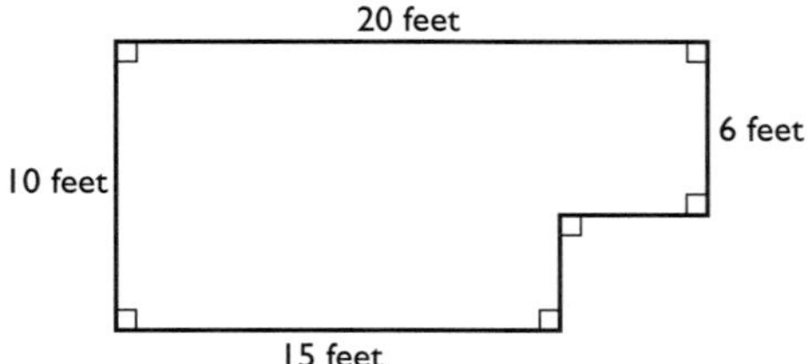

Determine the area to be covered with tiles. Each ceramic tile is one foot square. Speculate on the quantity that will be ordered.

Solution

Determine the missing dimensions and divide the floor into two rectangular shapes.

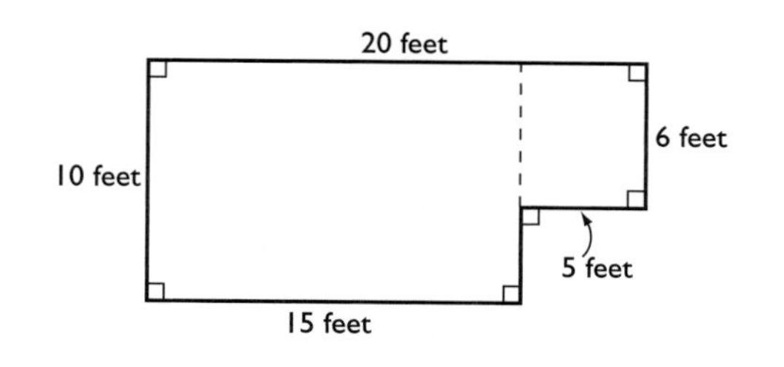

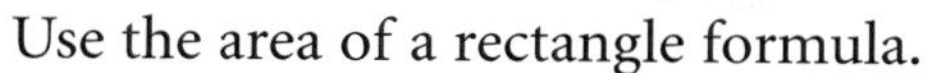

Use the area of a rectangle formula.

$A = lw$
$= 15 \times 10$
$= 150$

$A = lw$
$= 6 \times 5$
$= 30$

total area $= 150 + 30$
$= 180$

The area of the floor is 180 square feet. This suggests that 180 tiles would be needed. However, tiles will have grout between them so it is unlikely that exactly that many will be used. More may be ordered to allow for breakage and to have some spare tile to replace ones damaged in the future. As well, tiles may only be available in certain quantities because of how they are packaged.

Example 3

The living/dining room in Jeremy's condominium is rectangular with a semi-circular end that is all windows. Determine the floor area of the room.

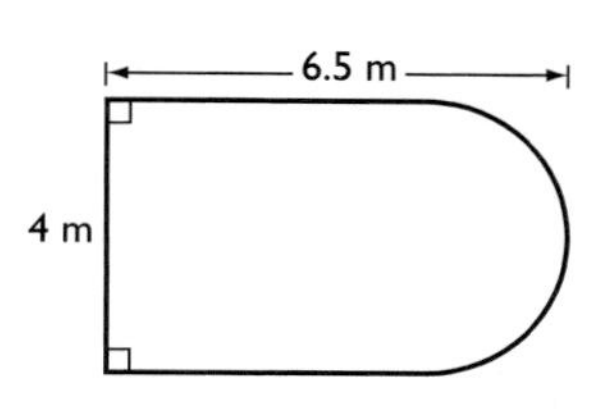

Solution

Determine the missing dimensions and divide the floor area into two shapes—a rectangle and a semi-circle.

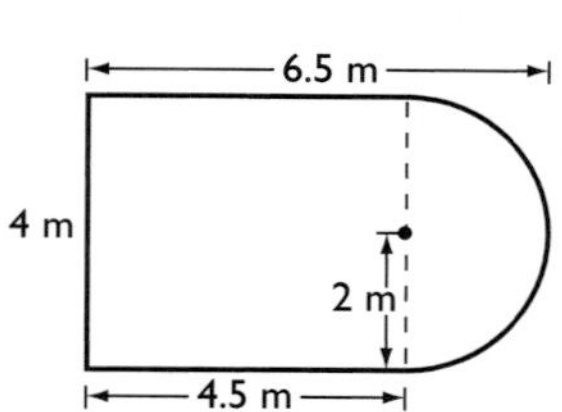

Use the area of a rectangle and the area of a circle formulas.

$A = lw$
$= 4.5 \times 4$
$= 18$

$\text{A} = \frac{1}{2} \text{ of } \pi r^2$
$\doteq \frac{1}{2} \times 3.14 \times 2^2$
$= 6.28$

total area $= 18 + 6.28$
$= 24.28$

The floor area of the room is 24.28 m^2.

Practise

1. Skills Check Match each formula with its description.

a) $P = 2l + 2w$
b) $C = \pi d$
c) $A = lw$
d) $A = \pi r^2$

i) area of a circle
ii) area of a rectangle
iii) perimeter of a rectangle
iv) perimeter or circumference of a circle

2. When mixing rug shampoo concentrate with water, you need to know the area to be cleaned. Calculate the area of each rug or carpeted floor.

a)
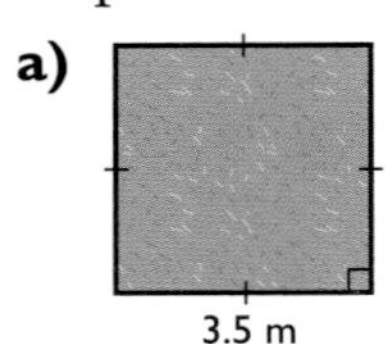

b)
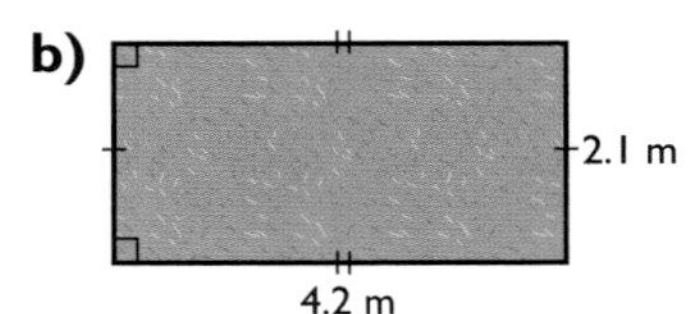

c)
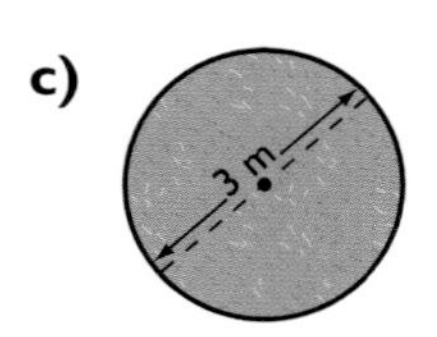

d)
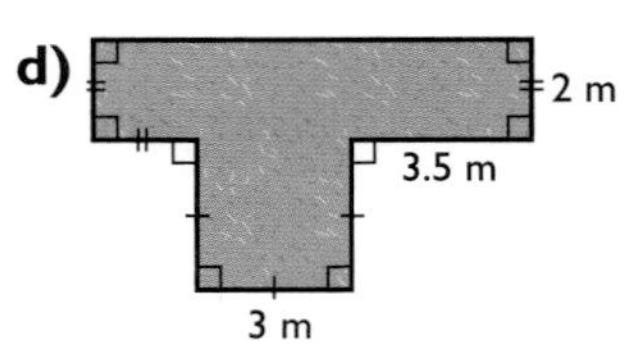

3. Calculate the amount of fencing or edging required around each garden.

a)
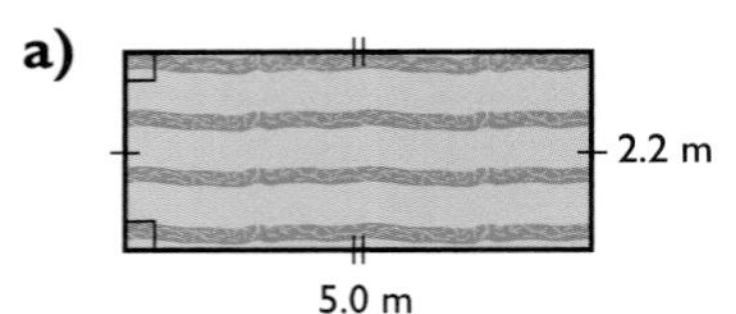

b)
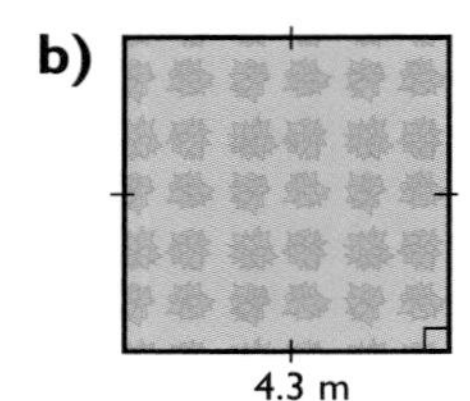

c)
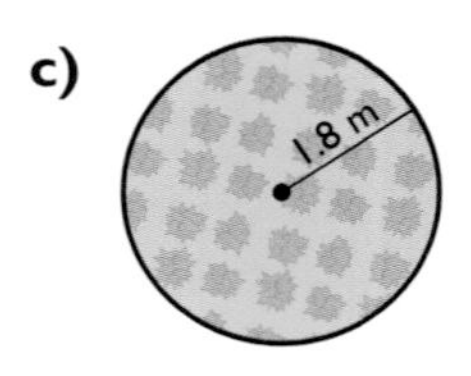

d)
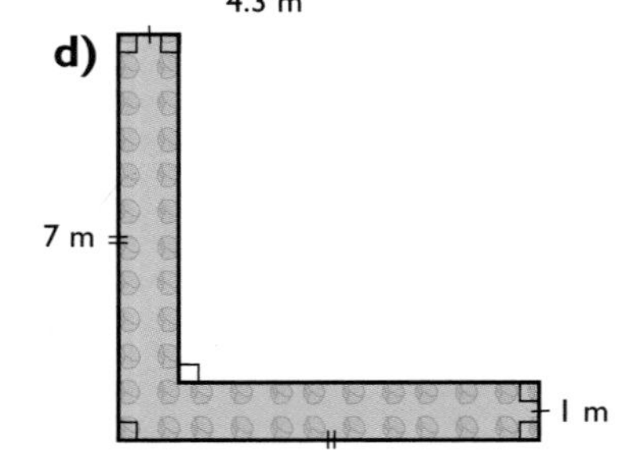

4. Determine the area of floor that is not covered by the round carpet.

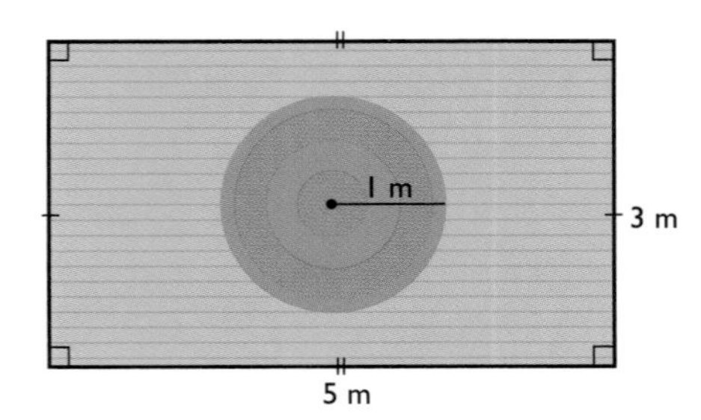

5. Amah works for a company that uses heavy road-building equipment. Amah's boss tells him to put a yellow warning tape around the perimeter of the asphalt-laying machine. What does Amah's boss mean?

6. Cheryl is buying paint for her rectangular living room. Two walls are 12 feet by 8 feet and the other two walls are 10 feet by 8 feet. There is one 4 foot by 6 foot window and one 7 foot by 4 foot rectangular archway.

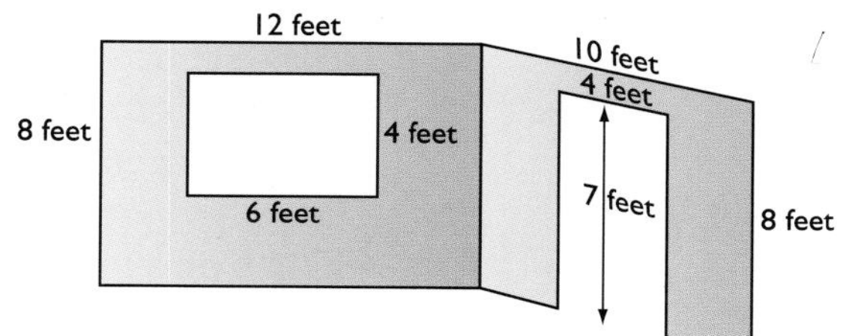

a) Calculate the area to be painted.

b) A can of paint covers 110 square feet. How many cans of paint should Cheryl buy to apply two coats of paint?

7. Steve is buying ceramic tile and baseboards for his sunroom floor. The room measures 12 feet by 9 feet. There are two 3 foot doorways.

a) Calculate the area of the floor.

b) Each tile covers one square foot and there are 12 tiles to a box. Steve plans to buys 5% extra to allow for breakage. How many boxes should he buy?

c) Calculate the perimeter of the floor.

d) Baseboards come in eight-foot lengths. How many pieces of baseboard should Steve buy?

8. A new house is built on a lot that measures 68 feet by 40 feet. The house measures 30 feet by 25 feet. It has a driveway measuring 15 feet by 12 feet.

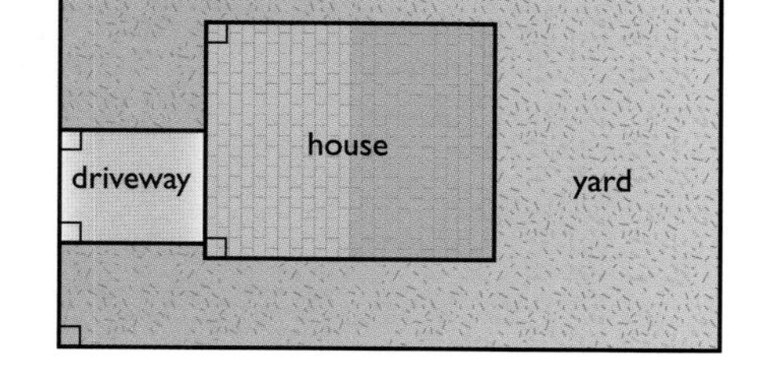

a) Describe how you would calculate the area of the yard.

b) Calculate the area of the yard.

c) Sod comes in strips that measure $1\frac{1}{2}$ feet by 6 feet. How many pieces of sod are required to cover the entire yard?

9. A racetrack has straight sides and semicircular ends. A railing is to be built around the outside edge of the racetrack.

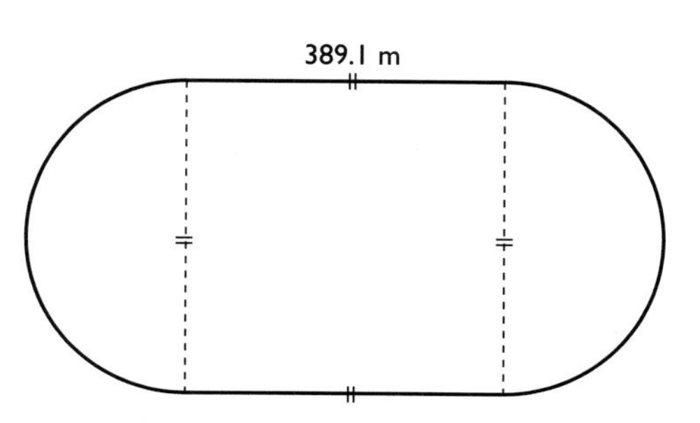

a) Determine the length of railing required, rounded to the nearest metre. What do you notice about the length?

b) If the railing costs $34.50/m, what will be the total cost of the railing?

8.3 Estimating Perimeter and Area

Explore

Your school council would like the school grounds to have new fencing and new ground cover. The council will begin the renewal process by getting a general idea of the costs, based on estimated dimensions. How could you estimate the perimeter of your school grounds? the area? What do you need to make the estimates?

Develop

1. a) Estimate the length and width of your classroom.
b) Estimate the perimeter of the classroom.
c) Estimate the area of the classroom.
d) Measure the length and width of the classroom.
e) Calculate the perimeter and area of the classroom.
f) How could you use your calculations to estimate the perimeter and the area of your school?
g) Compare your estimates and calculations from parts a) to e). Which estimates are closest to the corresponding calculations?
h) Discuss how you could improve your estimation techniques.

Practise

2. The school board is going to repave the school parking lot.
a) Estimate the area to be repaved. Explain the strategy you used.
b) If it costs $26.37/m^2 to repave, estimate the cost.

3. Ralph wants to open a small restaurant.
a) Estimate the floor area required for one table, measuring 1 m^2, and four chairs in Ralph's restaurant.
b) Estimate the floor area required for 10 sets of tables and chairs.
c) Ralph wants to have at least 10 sets of tables and chairs in his restaurant. What is the minimum floor area for the seating portion of his restaurant that Ralph must rent? How did you arrive at your answer?

4. You will be setting up a coffee shop in the school cafeteria for music night.
   a) Sketch the floor plan of the cafeteria.
   b) Estimate the number of sets of small tables and four chairs that you can put in the cafeteria.
   c) About how many people could be seated in the coffee shop?
   d) Explain the strategy you used to estimate the number of tables and chairs.
   e) How could you estimate the values more closely?

5. Estimate as closely as possible the perimeter and area of a room in your school or home.

6. Suppose you could refurnish your classroom to accommodate students working in groups. Instead of the current seating arrangement, there could be tables that measure 2 m by 1 m with six chairs at each.
   a) Sketch the floor plan of your classroom.
   b) Estimate the area required for each table and six chairs.
   c) About how many sets of tables and chairs could you fit in your classroom?
   d) Do you think your seating arrangement would work? Explain.

7. If sod costs $\$3.75/m^2$ laid, estimate the cost to lay sod on a playing field at your school.

8. Estimate the area of sidewalk along one side of one block near your home or school.

9. In everyday life, when do you think you will be working with perimeters and areas? Do you think you will need to estimate or calculate more frequently?

8.4 Career Focus: Flooring Installer

Anik is a flooring installer. She works for a company that installs a variety of indoor and outdoor floor coverings. Under the guidance of a skilled installer, she has learned how to install ceramic and other tiles, carpeting, and hardwood flooring while at work.

1. One job Anik had recently was laying ceramic tiles on a swimming pool deck.

a) Calculate the area that was tiled.

b) Each tile is one foot square and there are 12 tiles in a box. Anik allows 5% extra for breakage and plans to leave extra tiles with the customer. How many boxes of tiles would she have taken to the job?

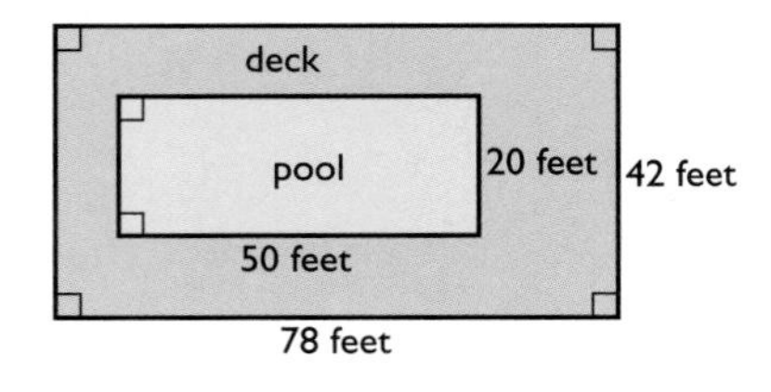

c) The cost of the tiles is $39 per box and when installed only GST applies. What was the cost to the customer of the tiles, not including other supplies and labour?

2. Another job Anik had recently was installing wall-to-wall carpeting in a living/dining room. A nailer strip was placed around the perimeter of the room to hold the carpeting in place against the walls.

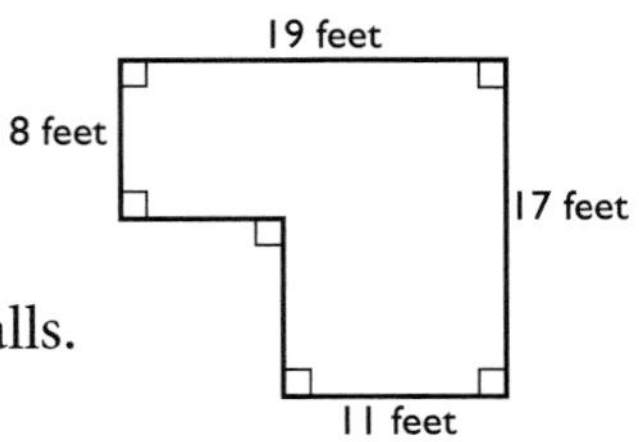

a) Calculate the perimeter of the room.

b) Nailer strips come in four-foot lengths. How many nailer strips were needed for the room?

c) Carpet is always laid running the same direction and with the fewest seams possible. It comes 12 feet wide. How many feet of carpet were needed for this job? How much carpet was left over?

8.5 Enlargements

Explore

What happens to the area of a rectangular photograph when you double the length and the width?

What happens to the photograph's area when you triple the length and the width?

Draw, use geometry software, or use concrete materials to check.

Develop

1. Lisa's grandmother liked a photograph that Lisa took of her younger brother and asked for a copy. Lisa made an enlargement for her grandmother. The enlargement is twice as long and twice as wide as the original photo, which is 5 inches by 7 inches. Calculate the area of the original and of the enlargement. Describe how much the area increases as the dimensions increase.

Practise

2. Michael has an old computer monitor with a screen that measures 11 inches by 9 inches. He buys a new monitor with a screen that measures 22 inches by 18 inches.

a) Describe the relationship between the dimensions of the old and new screens.

b) Calculate the areas of the old and new screens.

c) Select the relationship that compares the area of the new screen with the area of the old.

i) two times as large **ii)** four times as large **iii)** nine times as large

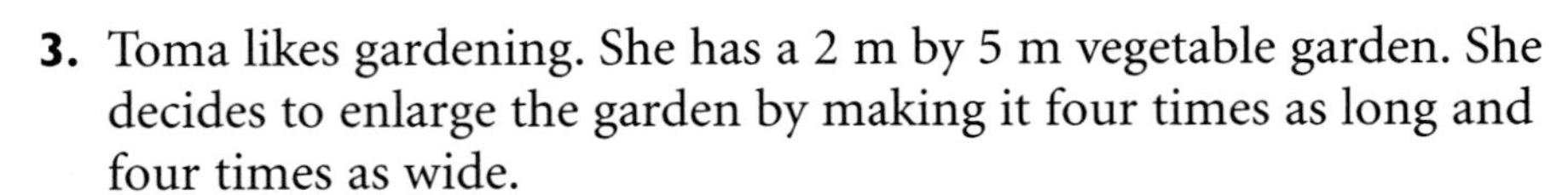

3. Toma likes gardening. She has a 2 m by 5 m vegetable garden. She decides to enlarge the garden by making it four times as long and four times as wide.

a) Find the length and width of the new vegetable garden.
b) How many times as large as the old vegetable garden is the new?

4. A portion of road map measures 8 cm by 7 cm. That portion is enlarged to 24 cm by 21 cm and more detail is shown.

a) How many times as long and wide as the original portion of the map is the enlarged portion?
b) How many times as large as the area of the original portion of the map is the area of the enlarged portion?
c) Check your answer to part b) by calculating the areas of the enlarged and original portions.

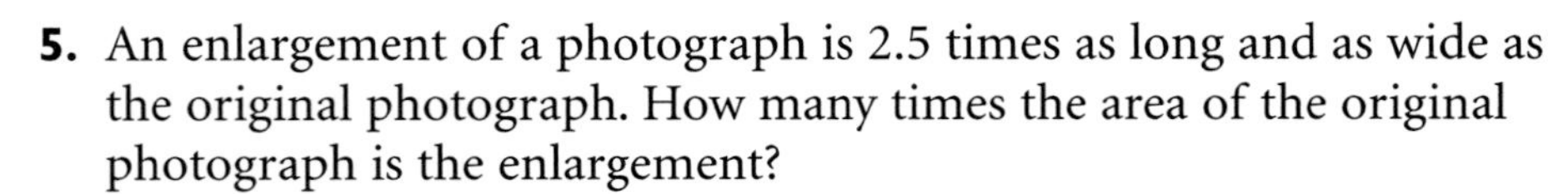

5. An enlargement of a photograph is 2.5 times as long and as wide as the original photograph. How many times the area of the original photograph is the enlargement?

6. Jordan is making a children's play area six times as long and six times as wide as it currently is. Determine how many times as large the area will be.

7. Show that if the dimensions of a TV screen are tripled, the area of the screen will be increased by a factor of nine.

8. Crystal wants to increase the area of the sign on her ice cream cart so that it is four times the current area. Determine how Crystal should change the length and width of her sign.

9. Recall that the area of a circle is found using the formula $A = \pi r^2$. A circular window has a diameter of 0.4 m.

a) Apply what you have learned to predict the relationship between the area of that window and the area of one with each radius.
i) 0.2 m **ii)** 0.1 m **iii)** 0.4 m
b) Calculate the areas to check your predictions.

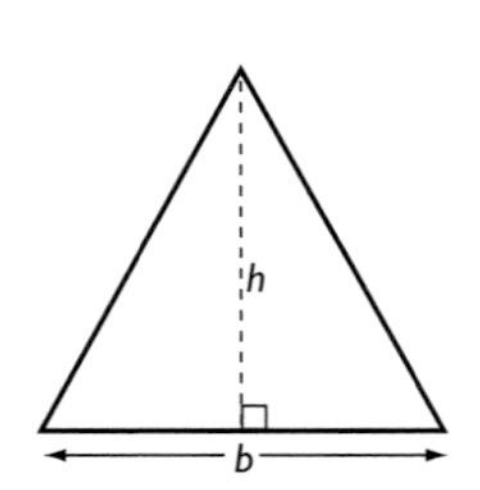

10. The area of a triangle is found using the formula $A = \frac{1}{2}bh$, where b and h are as shown.

a) Apply what you have learned to predict the relationship between the area of a triangular piece of stained glass art with a base of 15 cm and a height of 15 cm and one with the base and height doubled.
b) Calculate the areas to check your prediction.

8.6 Scale Drawings

Explore

Imagine that a sheet of $8\frac{1}{2}$" by 11" paper is the floor of a TV room that measures $8\frac{1}{2}$ feet by 11 feet. The TV room is in an apartment, a door is centred on one of the shorter walls, and there is no window or closet. What size of cut-outs should you make to model each of a TV, sofa, chair, table, and any other furniture you might want in the room? Make the cut-outs and arrange them to place the furniture to best suit the room.

Develop

Before a building is constructed, a **scale drawing** is usually made. Blueprints used by builders are detailed scale drawings. Floor plans, often used in selling new homes, are less detailed scale drawings. If a drawing is a scale drawing, it has measurements that are in proportion to those of the actual building. In the **Explore**, where an $8\frac{1}{2}$" by 11" sheet paper represented a floor that was $8\frac{1}{2}$ feet by 11 feet, the scale was 1 inch represents 1 foot.

Two-dimensional scale drawings show only one surface, for example, the front face of a building or the floor plan of a house.

1. Sal's variety store is 24 m by 18 m.

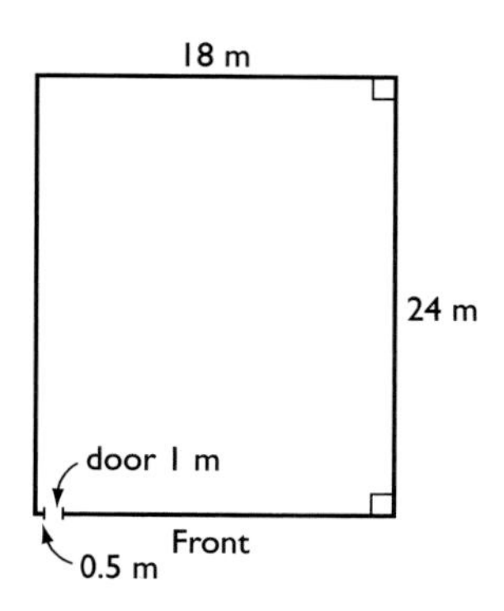

a) Use the scale 1 cm represents 1 m to outline the store floor on centimetre dot paper. Locate the door as shown.

b) There are shelves along the right wall of the store except the last 0.75 m at the back. There are shelves along the left wall of the store except the last 0.75 m at the back and the first 2 m at the front. All the shelves are 0.5 m deep. Mark the shelves on the scale drawing.

c) The refrigerated area of the store is along the back wall and measures 17 m by 0.75 m. It does not interfere with the side wall shelves. Mark the refrigerated area on the scale drawing.

d) The front counter, where the cash register is located, measures 0.5 m deep by 2 m long. It is parallel to the right wall and out 1.5 m from its shelves. The end of the counter is against the front wall of the store. Mark the counter on the scale drawing.

e) Sal wants to put in new free-standing shelving units. They are 6 m long and 1 m wide. Aisles between the shelves must be at least 1.5 m wide and the shelves should be parallel to the wall shelves. Using your floor plan, decide how many new shelving units Sal can use and where to put them.

Practise

2. **Skills Check** Complete each statement.
   a) If 1 m is represented by 1 cm, then 4.5 m would be represented by ___.
   b) If 1 m is represented by 1 cm, then 50 cm would be represented by ___.
   c) If 2 m are represented by 1 cm, then 1 m would be represented by ___.
   d) If 2 m are represented by 1 cm, then 6 m would be represented by ___.
   e) If 0.5 m is represented by 1 cm, then 1 m would be represented by ___.
   f) If 0.5 m is represented by 1 cm, then 4 m would be represented by ___.

3. **a)** Choose a suitable scale to create a scale drawing of the top of your desk on dot paper.
   b) Choose a suitable scale to create a scale drawing of your classroom on dot paper.
   c) Choose a suitable scale to create a scale drawing of an animal of your choice on dot paper.
   d) Explain how you chose a suitable scale for the scale drawings.

4. Create a scale drawing of your classroom either using dot paper and the scale you chose in question 3 b) or using design or drawing software and a suitable scale. Mark the doors, closets, and main pieces of furniture.

5. Choose a room you would like to create, such as a media or entertainment room, an exercise or workout room, or a sunroom. Create a scale drawing of the room either using dot paper or using design or drawing software and a suitable scale. Mark the doors, windows, furniture, and equipment.

8.7 Putting It All Together: Designing a Playground

A You are on a neighbourhood committee in a new subdivision that is giving the municipality input on what is needed for a playground for young children.

The committee has agreed that there should be a fenced circular wading pool; two sand boxes, one with dimensions twice those of the other; two swing sets and two slides, both designed for different age groups; at least three benches; some shrubs and flowers; and a garbage can.

Apply what you have learned in this chapter to do the following.

Decide on an area and shape of the playground that would accommodate the features specified and any others that you would like to include. Make a scale drawing using a suitable scale and mark all the features.

If the entire playground is fenced as well as the wading pool area, how much fencing will be required?

Identify at least three different right-angled corners, and explain how you could check that they are right angles.

8.8 Chapter Review

1. There are four support ropes on a tent like the one rope shown. What is the total length of the support ropes?

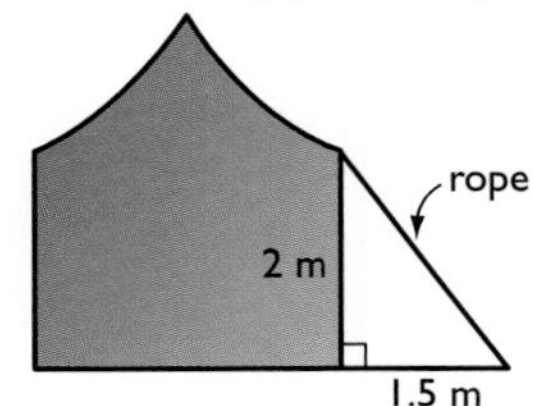

2. Mary is building a playhouse for her children. The floor will measure 6 feet by 8 feet. What distance should A to C measure so that there will be a right angle at B?

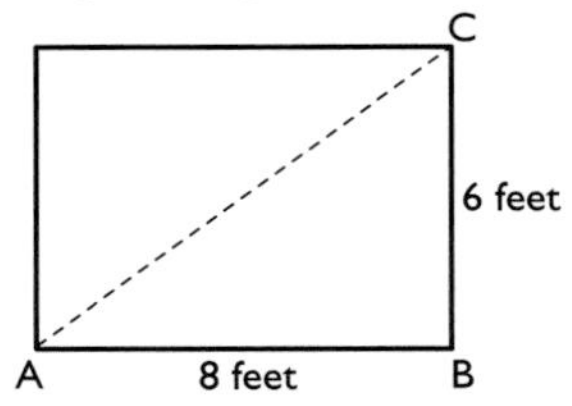

3. A 10 foot by 10 foot square eating area off a kitchen is getting ceramic tiles to match the kitchen and new baseboards. There is a four-foot opening to the kitchen and a four-foot opening to the hall.

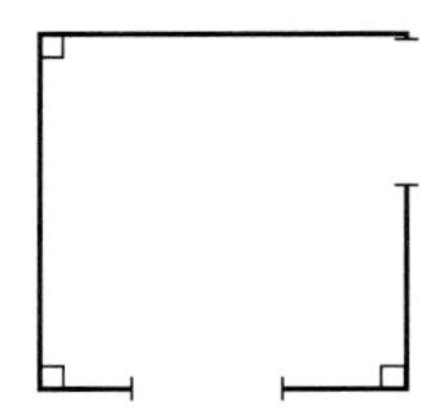

a) Calculate the area of the floor.
b) Each tile covers one square foot and there are 12 tiles to a box. Allowing 5% extra for breakage, how many boxes should be bought?
c) Calculate the perimeter of the floor.
d) Baseboards comes in eight-foot lengths. How many pieces of baseboard should be bought?

4. Calculate the outside and inside perimeter of the ring.

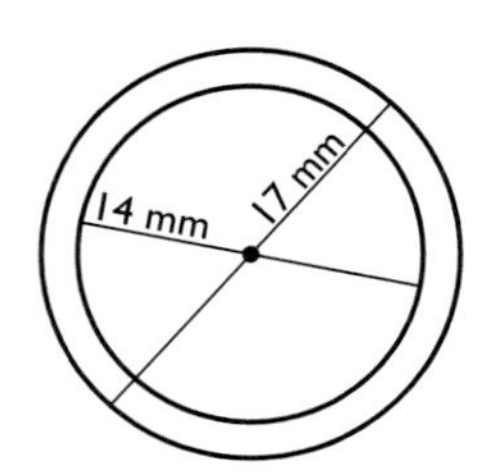

5. This combined living room and dining room has rugs as shown.

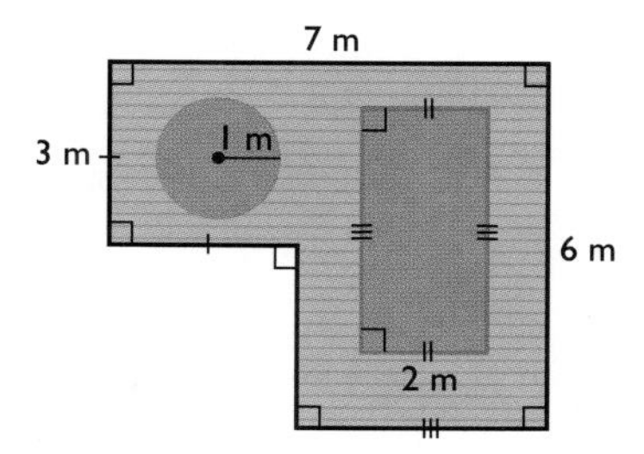

a) Calculate the floor area of the room.
b) Calculate the area of each rug.
c) Calculate the area of the uncovered floor.
d) The ceiling is the same shape as the floor. Calculate the length of crown moulding needed to go around the entire ceiling.

6. Explain how you could estimate the area of a park near your home.

7. An enlargement of a photograph is three times as long and three times as wide as the original photograph. How many times as large as the area of the original is the area of the enlargement?

8. A new amusement park sign is to be made with an area four times as great as the existing sign. Explain how the dimensions of the new sign will compare with the dimensions of the existing sign.

9. Make a scale drawing of a classroom wall, showing any doors, windows, cupboards, and blackboards on that wall, using a suitable scale.

10. A scale drawing of a shed floor measures 5 cm by 8 cm. The actual shed floor is to have dimensions 50 times as great as the dimensions of the scale drawing.

a) Make the scale drawing. Label the actual dimensions of the shed floor on the drawing.
b) What length should the diagonal of the actual shed floor be to ensure that the corners are right angles?

CHAPTER

9 Measurement and 3-D Design

In this chapter, you will

- estimate and calculate the volumes and surface areas of rectangular prisms
- estimate and calculate the volumes and surface areas of cylinders
- create three-dimensional drawings
- construct scale models

Near the end of the chapter, you will

- plan, design, and estimate the cost of a household improvement

9.1 Rectangular Prisms

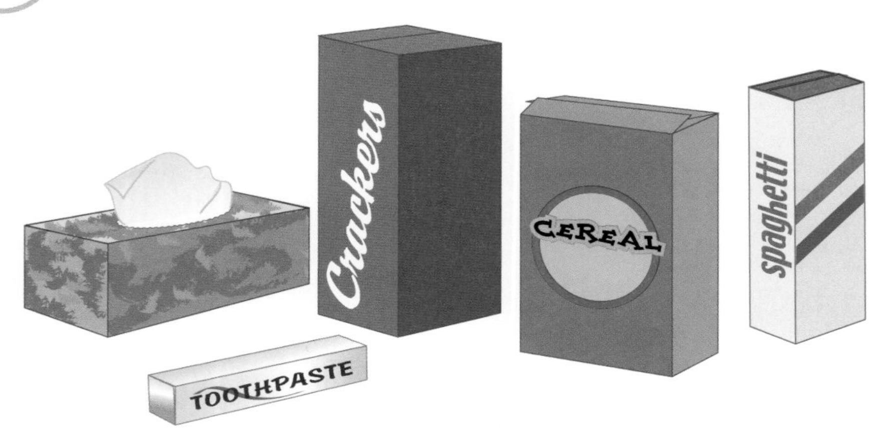

Explore

Examine some rectangular prism packages like the ones illustrated above.
What are the characteristics of a rectangular prism?
How could you calculate how much space a rectangular prism occupies?
How could you calculate the area of all the surfaces of a rectangular prism?

Develop

The amount of space that an object occupies is its **volume**. The sum of the areas of an object's faces is its **surface area**.

Remember that area is calculated by multiplying two dimensions, so the units are squared. Volume is calculated by multiplying three dimensions, so the units are cubed. For example, metres × metres × metres is cubic metres (m^3).

Example 1

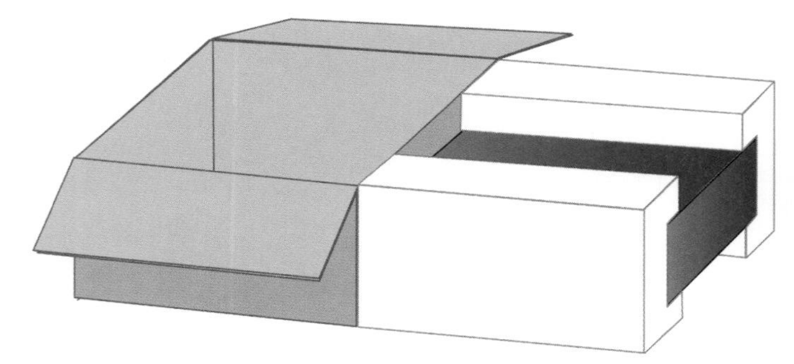

A DVD player is shaped like a rectangular prism with length 42.5 cm, width 30.0 cm, and height 12.5 cm. When packed in a cardboard box, it is protected by Styrofoam padding. The padding takes up 5.0 cm space all around the player. Estimate and calculate the volume that the cardboard box should be.

Solution

Volume of a rectangular prism can be determined by multiplying the area of the base by the height.

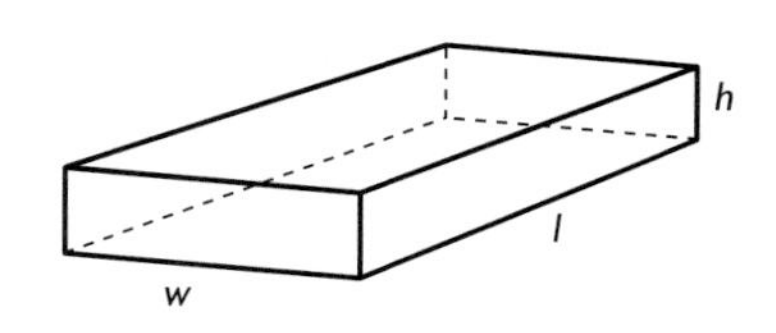

Volume = area of base × height

$= \text{length} \times \text{width} \times \text{height}$

$V = lwh$

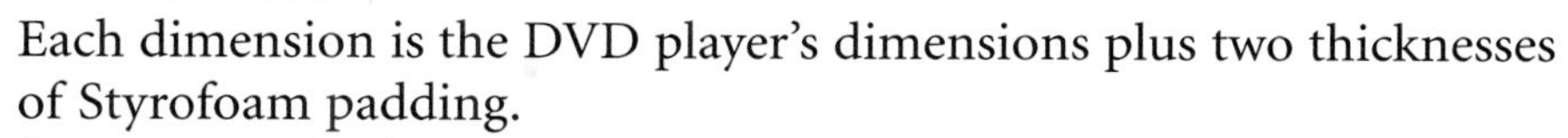

Each dimension is the DVD player's dimensions plus two thicknesses of Styrofoam padding.

$l = 42.5 + 2(5.0)$
$= 52.5$

$w = 30.0 + 10.0$
$= 40.0$

$h = 12.5 + 10.0$
$= 22.5$

Estimate.

$V = lwh$
$= 52.5 \times 40.0 \times 22.5$
$\doteq 50 \times 40 \times 25$
$= 2000 \times 25$
$= 50\ 000$

Calculate. 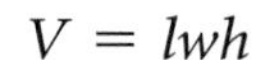

$V = lwh$
$= 52.5 \times 40.0 \times 22.5$
$= 47\ 250$

The volume that the cardboard box should be is 47 250 cm^3.

Example 2

Estimate and calculate the surface area of the DVD player's cardboard box from Example 1. Add 0.5 cm for the thickness of the cardboard to each dimension because the dimensions in Example 1 are inside measurements. Is this the amount of cardboard needed to make the box? Explain.

Solution

Surface area of an object is determined by adding the areas of all the faces. In a rectangular prism, opposite faces are congruent, that is, the same size and shape, so their areas are the same.

Surface area = 2(area of top/bottom face) + 2(area of front/back face) + 2(area of end face)

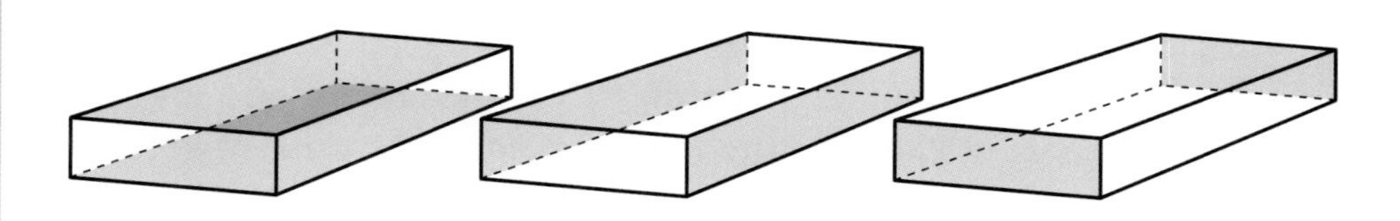

Outside dimensions

$l = 52.5 + 0.5$ $\quad w = 40.0 + 0.5 \quad h = 22.5 + 0.5$

$= 53.0 \quad\quad = 40.5 \quad\quad = 23.0$

Estimate.

$$\begin{aligned}\text{Surface area} &= 2(53.0 \times 40.5) + 2(53.0 \times 23.0) + 2(40.5 \times 23.0)\\ &\doteq 2(50 \times 40) + 2(50 \times 25) + 2(40 \times 25)\\ &= 2(2000) + 2(1250) + 2(1000)\\ &= 4000 + 2500 + 2000\\ &= 8500\end{aligned}$$

Calculate. 

$$\begin{aligned}\text{Surface area} &= 2(53.0 \times 40.5) + 2(53.0 \times 23.0) + 2(40.5 \times 23.0)\\ &= 4293 + 2438 + 1863\\ &= 8594\end{aligned}$$

The surface area of the DVD player's box is 8594 cm^2.
This would not be the amount of cardboard needed to make the box. A box has flaps where the cardboard overlaps, so more would be needed.

Practise

1. Estimate and then calculate the volume of each.

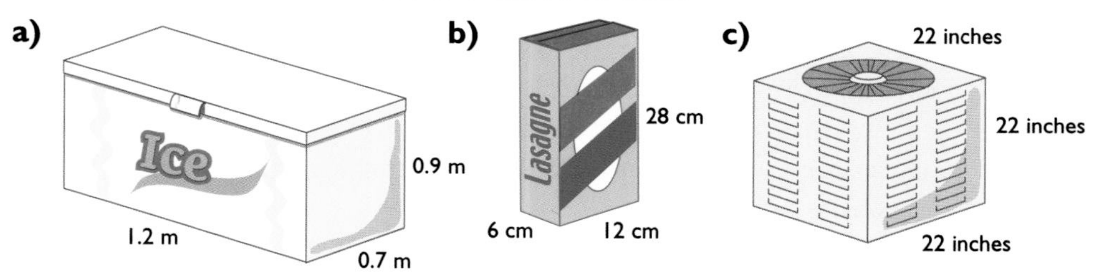

2. Estimate and then calculate the surface area of each rectangular prism in question 1.

3. A large rectangular basement room that measures 36 feet by 24 feet with a height of 7 feet 6 inches has had drywall installed and is ready for the walls, floor, and ceiling to be finished. Determine the following costs. Hint: Express all the dimensions in yards.
   a) carpet with underpad that costs \$20.25 square yard installed
   b) stucco on the ceiling that costs \$15.67 per square yard installed
   c) two coats of paint for the walls that comes in cans, where one can covers about 20 square yards and costs \$39.95

4. How could you estimate the volume of a rectangular building? Why would the volume of a building be of interest to someone?

5. A rectangular flower bed measures 3 yards by 2 yards. The bed must be dug out and filled with topsoil to a depth, or height, of $\frac{1}{4}$ yard. Topsoil is delivered for \$72 per cubic yard.
   a) Estimate and calculate the volume of topsoil required to fill the flower bed.
   b) Calculate the cost of the topsoil delivered.

6. Measure the dimensions and then calculate the volume of each.
   a) your classroom
   b) the inside of a drawer
   c) a tissue box
   d) a box of your choice

7. Stacks of 500 sheets of paper are wrapped in paper. A 500-sheet stack is $8\frac{1}{2}$ inches by 11 inches by 2 inches. The wrapping paper is $22\frac{1}{2}$ inches by 15 inches.
   a) Calculate the surface area of a 500-sheet stack.
   b) Calculate the area of the wrapping paper.
   c) How much wrapping paper is used in overlap? Explain.
   d) What is the area used in overlap as a percent of the surface area of the stack?

8. a) Measure the dimensions of a box, such as a toothpaste tube box.
   b) Calculate the surface area of the box.
   c) Disassemble the box and calculate the area of the cardboard used in the box.
   d) How much cardboard is used in overlap? Explain.
   e) What is the area used in overlap as a percent of the surface area of the box?
   f) How does the percent in part e) compare with the percent you found in question 7 d)?

9. A CD jewel case has dimensions 13.7 cm by 12.5 cm by 1 cm.

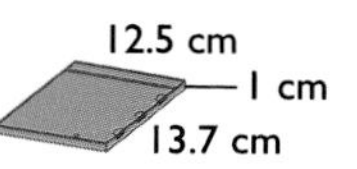

   a) Calculate the volume of 24 CD jewel cases.
   b) Calculate the surface area when 24 cases are stacked as shown.

i)

ii)
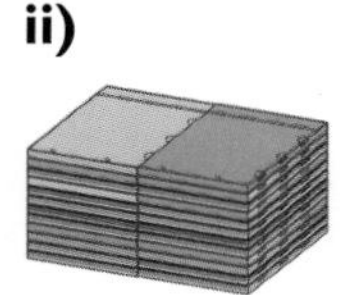

iii)
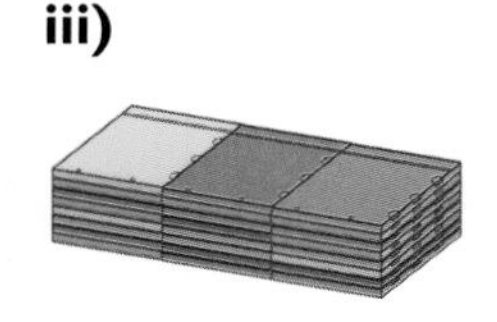

iv)
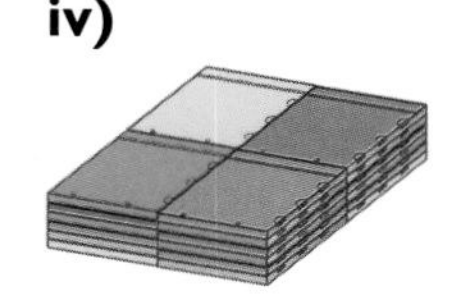

   c) Based on your findings in part b), predict which container would have the least surface area. Then, calculate each surface area to check.

i)
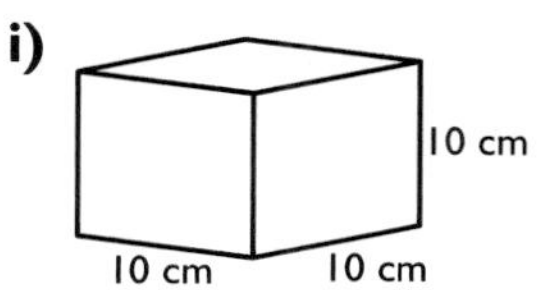

ii)
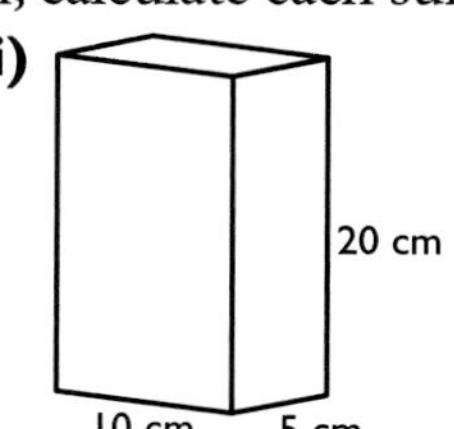

9.2 Cylinders

Explore

Examine some cylindrical cans like the ones illustrated above.
What are the characteristics of a cylinder?
How is a cylinder like a rectangular prism?
How is a cylinder different from a rectangular prism?
How could you calculate how much space a cylinder occupies?
How could you calculate the area of all the surfaces of a cylinder?

Develop

1. Examine a can with a label that can be cut off.

a) The volume of a rectangular prism can be determined using Volume = area of base × height. Could this formula be used to find the volume of a cylinder? Explain.

b) What shape is the base of a cylinder? What is the area formula for that shape?

c) Because the area of the base is length × width, the volume of a rectangular prism formula can be expressed as $V = lwh$. Express the volume of a cylinder in a similar way.

2. Surface area is the sum of the areas of all the faces.

a) How many faces does a cylinder have?

b) What is the shape of two of those faces? What is the area formula for that shape?

c) To determine the shape of the curved face, cut the label off in a straight line from top to bottom. What shape is the label when it is flattened? What is the area formula for that shape?

d) The distance around the circular base is the length of the curved face flattened and the height of the can is its width. Which formula would give the area of the curved face?

i) $lw = \pi r^2 h$ **ii)** $lw = \pi dh$ **iii)** $lw = \pi rh$

e) What is the formula for the surface area (all three faces) of a cylinder?

Example 1

A cylindrical hassock has a radius of 21 cm and a height of 38 cm. Estimate and calculate the volume of material in the hassock.

Solution

The volume of a cylinder is determined using
Volume = area of circular base × height.

$V = \pi r^2 h$

Estimate.

$V = \pi r^2 h$
$= 3.14 \times 21^2 \times 38$
$\doteq 3 \times 20^2 \times 40$
$= 3 \times 400 \times 40$
$= 48\ 000$

Calculate.

$V = \pi r^2 h$
$= 3.14 \times 21^2 \times 38$
$= 52\ 620.12$

The volume of the material in the hassock is 52 620.12 cm^3.

Example 2

The outside of the hassock in Example 1, including the bottom, is leather. Estimate and calculate the surface area of the hassock. Is this the amount of leather needed to make the hassock? Explain.

Solution

The surface area of a cylinder is determined using
Surface area = 2(area of circular bases) + area of rectangular face.
Surface area = $2(\pi r^2) + \pi dh$ where πd is the circumference of a circular base, and thus the length of the rectangular face; and h is the height of the cylinder, and thus the width of the rectangular face

Estimate.

Surface area $= 2(\pi r^2) + \pi dh$
$= 2 \times 3.14 \times 21^2 + 3.14 \times 42 \times 38$
$\doteq 2 \times 3 \times 20^2 + 3 \times 40 \times 40$
$= 2400 + 4800$
$= 7200$

Calculate.

Surface area $= 2(\pi r^2) + \pi dh$

$= 2 \times 3.14 \times 21^2 + 3.14 \times 42 \times 38$

$= 7780.92$

The surface area of the hassock is 7780.92 cm^2.

This would not be the amount of leather needed to make the hassock. There would need to be more leather for overlap where there are seams.

3. a) Measure the diameter and height of a can to the nearest tenth of a centimetre.

b) Calculate the volume of the can.

c) Compare your calculated volume of the can with the capacity given on the can.

d) 1 cm^3 holds 1 mL. When measuring dimensions of packages with capacities given on their labels and then calculating their volume, how can you tell if your calculations are reasonable?

Practise

4. Estimate and then calculate the volume of each can.

a) 8.5 cm, 11 cm

b) 10.7 cm, 17 cm

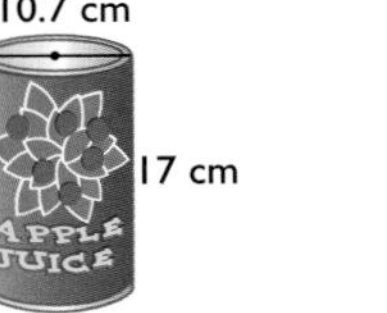

c) 9.5 cm, 21.5 cm

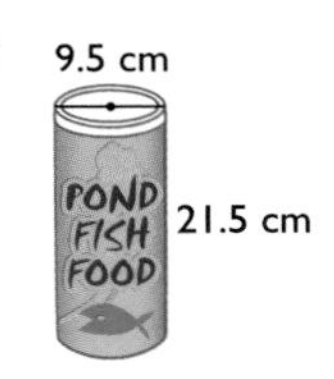

5. Estimate and then calculate the surface area of each can in question 4.

6. Janelle and Siraj are building a water garden. The pond is circular and 1.5 m in diameter. It will be filled with water to a depth, or height, of 0.5 m.

a) Calculate the volume of water in the pond to the nearest hundredth of a metre.

b) Janelle and Siraj plan to stock the pond with goldfish. One goldfish requires about 0.125 m^3 of living space. About how many goldfish can Janelle and Siraj put in the pond?

c) The pond requires a liner that must cover the bottom and the side. Calculate the cost of the liner if the charge is $14.76/$m^2$.

7. A cylindrical drinking glass is 9 cm in height by 7 cm in diameter. Will it hold 250 mL of liquid?

8. A cylindrical coffee mug measured from the inside is 9 cm in height by 7 cm in diameter. About how many millilitres of liquid will it hold?

9. Naomi makes pillar candles.
   a) Which requires more wax—three regular candles, each with a diameter of 3 inches and a height of 6 inches or a three-wick candle with a diameter of 6 inches and a height of 6 inches?
   b) If Naomi sells a regular candle for \$6, suggest a reasonable selling price for a three-wick candle. What would affect the price other than the amount of wax used?

10. Andy produces cylindrical glass bottles for perfume. Each bottle must hold 100 mL of perfume. The glass from which the bottles are produced is very expensive. The lids to fit any of the sizes below cost the same to produce.
    a) Verify that each size below will hold 100 mL.
       i) $r = 1.5$ cm, $h = 14.2$ cm
       ii) $r = 2.2$ cm, $h = 6.6$ cm
       iii) $r = 2.5$ cm, $h = 5.1$ cm
    b) Andy wants to minimize production costs by using the least glass needed. Which size should he choose for the bottle? Why?

11. A cylindrical tube can be made by rolling a sheet of $8\frac{1}{2}$" by 11" paper until the edges just meet.
    a) Predict how the volumes would compare in these two instances: when the sheet of paper is rolled lengthwise to make a long thin tube and when it is rolled the other way to make a short wide tube.
    b) Calculate the volume of each to check your prediction.
    c) What about the volume of a cylinder formula would suggest the results you found in part b)?

12. "Round" bales of hay are cylindrical with a diameter of 5 feet and a length, or height, of 4 feet. Small "square" bales of hay are rectangular prisms measuring 38 inches by 16 inches by 14 inches. How many square bales would be needed to feed as many animals as one round bale? Explain.

9.3 3-D Drawings

Two-dimensional scale drawings, as you saw in the previous chapter, show only one surface. Drawings can be made to show three dimensions as well. You have seen such drawings of boxes and cans in the previous sections of this chapter.

Explore

Make a three-dimensional drawing of the interior of an open box, such as a shoe box, using dot paper, using geometry, design, or drawing software, or drawing freehand.

Develop

For questions 1 to 4, use dot paper, use geometry, design, or drawing software, or draw freehand.

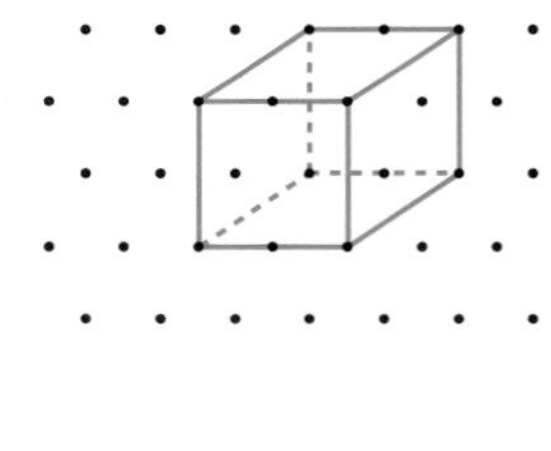

1. Make three-dimensional drawings of the following.

a) a rectangular prism that is a cube
b) a cylinder that is wider than it is tall
c) the inside of a classroom with the teacher's desk in view, showing three walls

Practise

2. For each aquarium, make a three-dimensional drawing.

a) an aquarium that has a square base and is about as wide as it is deep
b) a cylindrical aquarium that is about three times as wide as it is deep
c) an aquarium that has a square base and is about three times as wide as it is deep
d) a cylindrical aquarium that is about three times as deep as it is wide

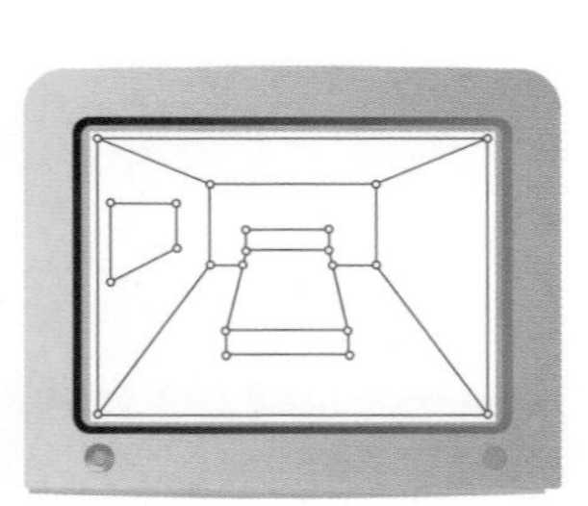

3. Make a three-dimensional drawing of a room in your home, showing three walls, any doors or windows in those walls, and a piece of furniture.

4. Decide upon a room that you would like to create, or use the room that you created in Section 8.6 and made a scale model floor plan for. Make a three-dimensional drawing of the room, showing three walls, any doors or windows in those walls, and several pieces of furniture.

9.4 Scale Models

A **scale model** is a larger or smaller representation of an object where the measurements are in proportion to the actual object.

Architects and designers use scale models to represent buildings, bridges, and other structures to help clients visualize the final structures. Set designers use scale models for theatre productions.

Explore

Use pipe cleaners to make a model of a chair. Determine the scale you used.

Develop

To construct a scale model, you

- determine the amount of space and materials with which you have to work
- determine the measurements of the actual object
- determine an appropriate scale to use
- convert actual measurements to scale measurements
- construct the model

1. Furniture for a doll house is made using the scale 1:24, or 1 inch represents 24 inches or 2 feet. What is the actual length of each item?

a) a table $2\frac{1}{2}$ inches long **b)** a sofa $3\frac{3}{4}$ inches long

2. What would be a suitable scale to use to make a model of each of the following if you wanted to be able to hold the model in your hand?

a) a chair that is 1 m high
b) a soldier that is 6 feet tall
c) a fire truck that is 15 m long

Practise

3. The CN Tower has a base with a diameter of 66.6 m and a height of 533.33 m. If you build a scale model where the scale is 1:50, or 1 m represents 50 m, the base on the model would have a diameter of 1.33 m. What height of room would you need for displaying the model?

4. Jacob wants to create a scale model of the Parliament Buildings in Ottawa for a class project. His research reveals these measurements.

- The Centre Block, excluding the Peace Tower, has exterior dimensions of 145 m in width by 75 m in height.
- The height of the Peace Tower from the ground to the base of the flagpole is 55 m.
- The height from the ground to the centre of the clock face is 65 m.
- The diameter of the clock face is 5 m. The length of the minute hand is 2.5 m. The length of the hour hand is 1.5 m.
- The flag measures 4.5 m by 2.0 m.

a) Jacob decides on a scale of 1:100 for his model. Determine the scale measurements for his model in centimetres.

b) Jacob needs to carry his model to school. Should he reconsider his scale? Why or why not?

5. A diesel engine is shaped much like a rectangular prism with dimensions about 12.5 m long by 1.5 m wide by 3.5 m high. The engine has 10 wheels, 5 along each side, each with a diameter of 1 m.

a) Which of the following is an appropriate scale for a model of this diesel engine?

i) 1 cm represents 1 m

ii) 1 cm represents 50 cm

iii) 1 cm represents 2 m

b) Convert the actual measurements of the engine to model measurements.

c) Made a three-dimensional sketch of the model.

d) Construct a scale model of the diesel engine.

e) Calculate the volumes of the actual engine and the model. Compare them.

f) Calculate the surface areas of the actual engine and the model. Compare them.

6. Choose a structure that interests you, such as a building, a garden, or a bridge. Construct a scale model of the structure. Provide all necessary dimensions and your chosen scale.

9.5 Career Focus: Hobby Store Clerk

Tung works in a hobby store that specializes in model cars, trains, airplanes, and spacecraft. While working there, he has learned a great deal about the many different scales used in modeling.

1. How should Tung reply to a customer who asks, "What does scale mean?"
2. How should Tung reply to a customer who asks, "What does the 1:24 on the model box mean?"
3. A representation of a scene with 3-D models is called a **diorama**. Why would Tung recommend that all of the models in a diorama have the same scale?
4. Most modeling scales trace back to scales used for architectural drawings and models. The most commonly used scales tend to be ratios that permit easy conversions in the imperial measurement system. For example, a scale of 1:12 where 1 inch represents 12 inches or 1 foot and a scale of 1:16 where $\frac{1}{16}$ of an inch represents 1 inch are common. These two scales and their multiples form the basis of most modeling scales. Multiples of 1:12 include 1:24 and 1:36, and multiples of 1:16 include 1:32 and 1:48.
   a) Explain why the following are standard modeling scales found in the hobby store where Tung works.
      i) 1:24, the scale for most cars
      ii) 1:72, a common scale for airplanes
      iii) 1:96, a typical scale for spacecraft
   b) Explain why the scales 1:25 and 1:43 are considered to be "oddball" by most hobbyists.
   c) What types of scale are appropriate for working in the metric measuring system? Why?

9.6 Putting It All Together: Finishing a Basement

Your task is to plan, design, and estimate the cost of finishing a 10 m by 8 m basement in a home. You may choose to create a workshop, a recreation room, a bedroom, a bathroom, or any combination of these rooms. The outside walls are framed and insulated. The floor plan to the right shows the position of the stairs, the furnace, the windows, the support posts, and where plumbing is roughed in for a bathroom.

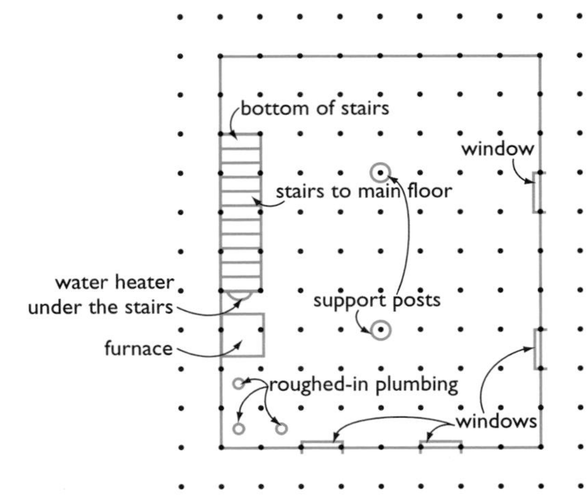

Apply what you have learned about measurement and design in both two dimensions and three dimensions to do the following.

Make a floor plan scale drawing of the basement on grid paper. Indicate the scale used. Decide on the position of any partitions and doorways. Indicate the dimensions of each room created and the purpose of each room.

Estimate the cost, excluding labour, of
- finished walls, floor, and ceiling
- any doors or cupboards
- any electrical and plumbing fixtures

Assume that labour is twice as much as the materials and estimate the labour costs.

Estimate the cost of furniture, electronics, or appliances for your room(s).

What is the total cost per square metre of your completed project?

Make a three-dimensional drawing or a scale model of one room.

9.7 Chapter Review

1. What is the meaning of each term? Sketch and label a diagram to help explain each.

a) volume **b)** surface area **c)** scale

2. Which container has the greater volume? How much greater?

a)

b)

3. Which container in question 2 has the greater surface area? How much greater?

4. Make a three-dimensional diagram of a bedroom with a bed, desk, and chest of drawers.

5. A model car is constructed using a scale of 1:24.

a) Is this an enlargement or a reduction of the actual object?

b) A measurement on the model of the car is 2 inches. What is the measurement on the actual car?

c) What would the model measurement be for an actual measurement of $3\frac{1}{2}$ feet on the car?

6. Grace works for an architectural company that is planning a new building that will be shaped like a rectangular prism. The building, with three floors, will measure 50 m by 80 m by 12 m in height. A cylindrical greenhouse with a diameter of 7 m and a height of 3.5 m will be constructed on the roof.

Grace must gather the following information:

- the amount of space in the building, including the greenhouse
- the area that the building will occupy on the property
- the total cost of brick blocks, which cost $\$10.48/m^2$, for the outside walls of the building, not including the greenhouse, assuming 50% of the outside walls will be windows and doors
- the total cost of glass, which costs $\$7.98/m^2$, only for the greenhouse walls and roof
- an appropriate scale for a scale model of the building, as well as the scale dimensions

Determine the measures and costs, and create a report for Grace.

CHAPTER

10 Transformations and Design

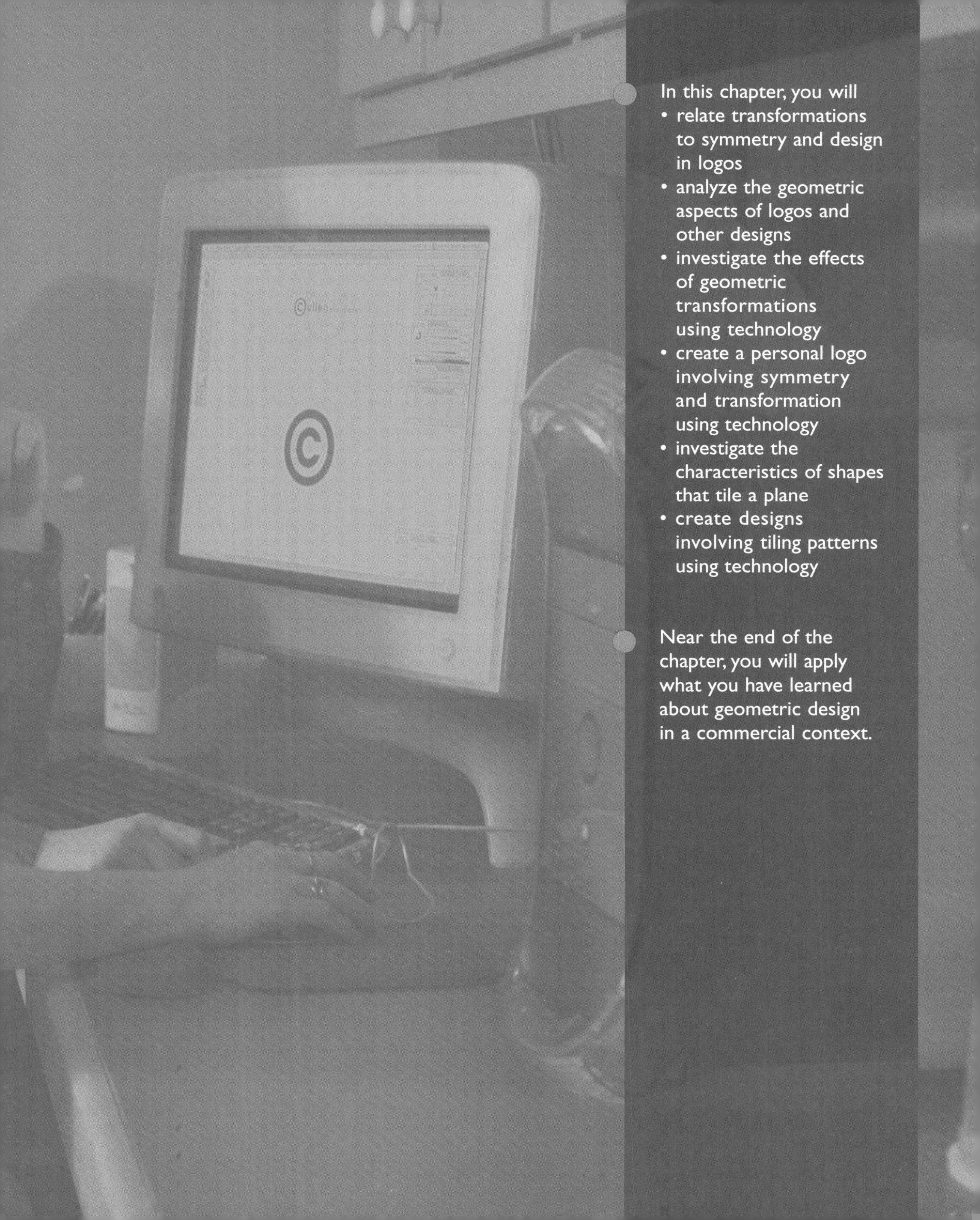

In this chapter, you will

- relate transformations to symmetry and design in logos
- analyze the geometric aspects of logos and other designs
- investigate the effects of geometric transformations using technology
- create a personal logo involving symmetry and transformation using technology
- investigate the characteristics of shapes that tile a plane
- create designs involving tiling patterns using technology

Near the end of the chapter, you will apply what you have learned about geometric design in a commercial context.

10.1 Geometric Aspects of Design

Businesses, industries, and government agencies often use **logos**, or distinctive symbols, to promote their images. Most people understand and relate to visuals much more easily than they do to words or text. A logo should be simple and unique, and have visual appeal. Once a logo is established, people will readily recognize it. For example, the red cross is a logo for medical assistance that is known around the world.

Explore

Examine the logos that appear below. Describe how symmetry and geometric transformations have been used. What do you think makes these logos appealing and appropriate for the organizations they represent?

Example 1

Two types of symmetry exist—line, or reflectional, symmetry and turn, or rotational, symmetry. Which type is shown in each flag? For line symmetry, locate the line of symmetry and for turn symmetry, locate the turn centre.

a) Canada

b) Anguilla

c) Czech Republic

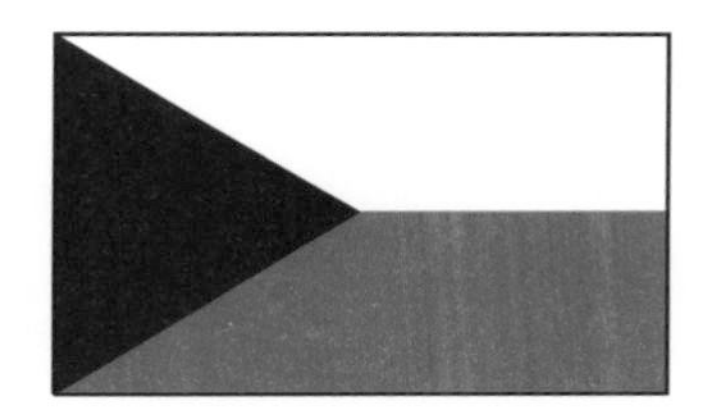

d) Morocco

Solution

a) line

b) turn

c) line

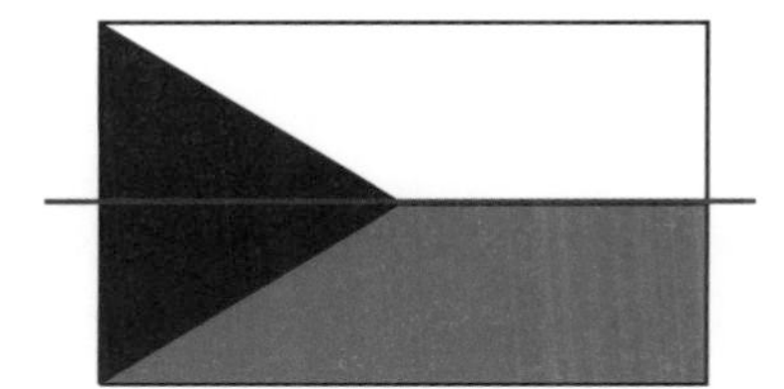

d) turn

Example 2

Four geometric transformations are shown in the designs below. Name each transformation and identify the design that shows it.

a) This transformation has the effect of sliding an object horizontally and/or vertically. The image is the same size and shape, and has the same orientation as the original.
b) This transformation has the effect of moving an object in a circular motion about a point. The point is the centre for turn symmetry. The image is the same size and shape, but has a different orientation than the original.
c) This transformation has the effect of creating a mirror image of the object about a line. The line is a line of symmetry. The image is the same size and shape, but has the opposite orientation as the original.
d) This transformation has the effect of enlarging or reducing an object. The image is the same shape and has the same orientation, but is not the same size as the original.

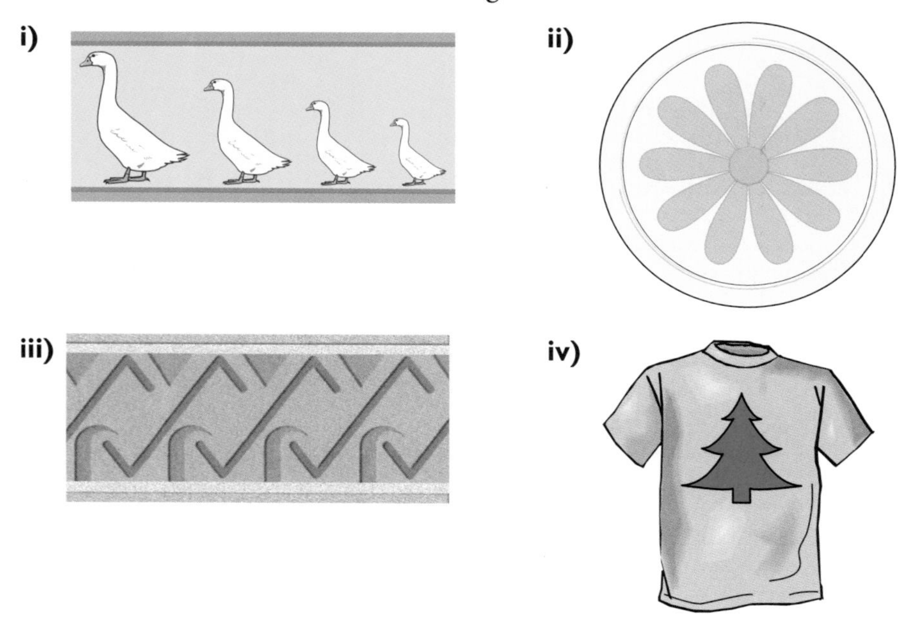

Solution

a) translation, iii) **b)** rotation, ii) **c)** reflection, iv) **d)** dilatation, i)

Logos usually contain graphic symbols. Sometimes, these are created using geometric transformations. Geometric transformations change the positions of shapes and thus their appearance. The symbols that form the basis of a logo may undergo translations, rotations, reflections, and dilatations.

Example 3

Analyze the geometric aspects of each logo.

a) Halton District School Board **b)** Waterloo District School Board

Solution

a) The stylized H in the Halton District School Board logo has line symmetry about a vertical line through the centre of the H. The stylized H shows a reflection.

b) There is no symmetry in the Waterloo District School Board logo. The stylized people show a dilatation, the one on the right being a reduction of the one on the left. The circles show a dilatation, the inner one being a reduction of the outer one and centred on it.

Practise

1. Analyze the geometric aspects of each traffic sign.

2. Analyze the geometric aspects of each school board logo.

a) District School Board of Niagara

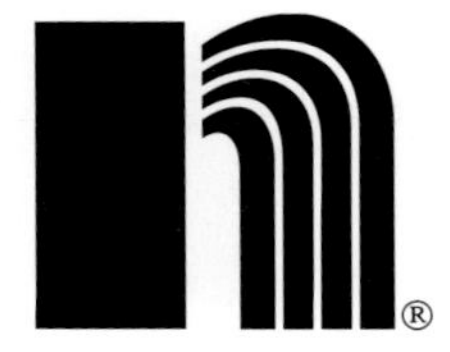

b) Halton Catholic District School Board

c) Near North District School Board

d) Algonquin & Lakeshore Catholic District School Board

3. How can you easily recognize each transformation?
   a) translation
   b) reflection
   c) rotation
   d) dilatation

4. a) Analyze the geometric aspects of this wallpaper design.

   b) Sketch a wallpaper design, and describe how you used symmetry and/or transformations.

5. a) Use print sources, such as newspapers and magazines, or the Internet to locate logos and other designs that show symmetry and/or transformations. For each, describe how symmetry or transformations have been used.
   b) What geometric aspects do you think contribute to the appeal of the designs that you examined in part a)? Explain.

6. Symmetry and transformations appear in many objects around you. Consider woodwork on furniture, dinnerware patterns, wallpaper patterns, decorative trim on buildings, and car marques.

   Identify five or more designs in your surroundings that involve symmetry and/or transformations. For each, describe the use of symmetry or transformations.

10.2 Investigating Design Using Technology

In this section, you will investigate various geometric designs that you can create using one of your initials. You may decide to use the results to design your own personal logo.

Explore

Explore geometric transformations using dynamic geometry software. Create a simple shape and investigate the effects of various translations, rotations, reflections, and dilatations. Describe the effect of each transformation on the original shape.

Develop

1. What are the initials from your name? Describe any symmetry in the letters. Choose one initial to construct and transform.
2. In the dynamic geometry software, choose appropriate settings to not show labels automatically, to measure in centimetres, and to have grid and axes showing. The grid will help you analyze the effect of a transformation.
3. Construct your initial with one point at the origin, or where the axes intersect. If necessary, use line segments instead of curves for easier construction and transformation.
4. Select your initial. Transform it using a translation by a rectangular vector. Choose a horizontal translation of 2 and a vertical translation of 3. This transformation can be written (2, 3). Describe the effect of the translation on the original initial. Save the file. Hide the transformation so that just your original initial is showing on the axes and grid.

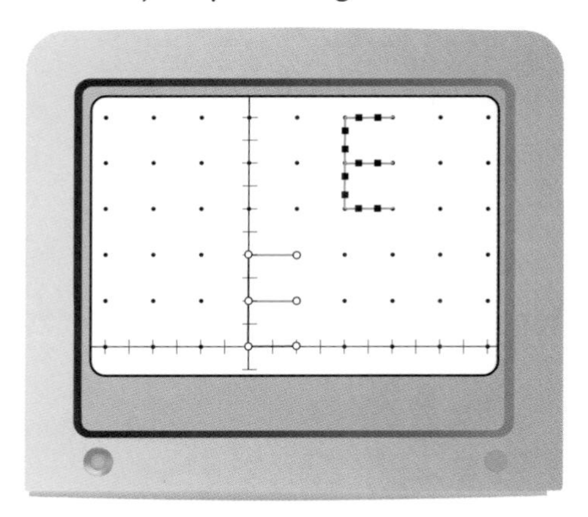

5. Repeat question 4 for each of the following translations. Save each translation in a separate file. Describe the effect of each translation.

a) (0, 3) **b)** (3, 0) **c)** (0, –3) **d)** (–3, 0)
e) (3, 3) **f)** (–3, –3) **g)** (–3, 3) **h)** (3, –3)

6. a) Hide the last transformation so that just your original initial is showing on the axes and grid.

b) Use the line tool, not the line segment tool, to construct a line through your initial. If your initial is symmetrical, construct a line that is not the line of symmetry. Mark this line as a mirror.

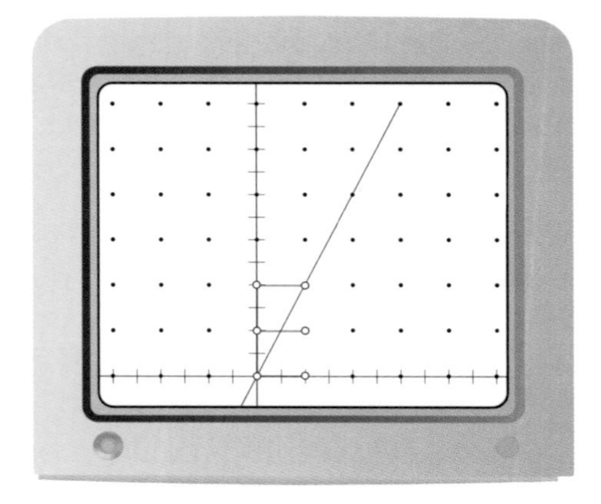

c) Select your initial. Transform it using a reflection. Describe the effect of the reflection on the original initial. Describe any symmetry that you see. Save this file.

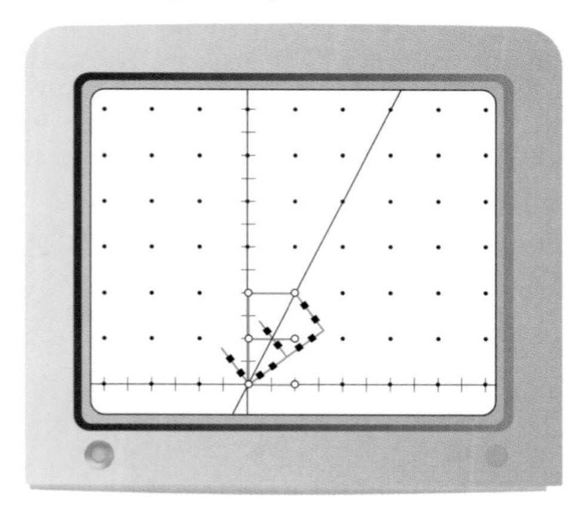

7. Repeat question 6, but construct a line outside your object. Describe differences and similarities between the reflection images in questions 6 and 7.

8. **Skills Check** Match each angle with the correct degree measure.

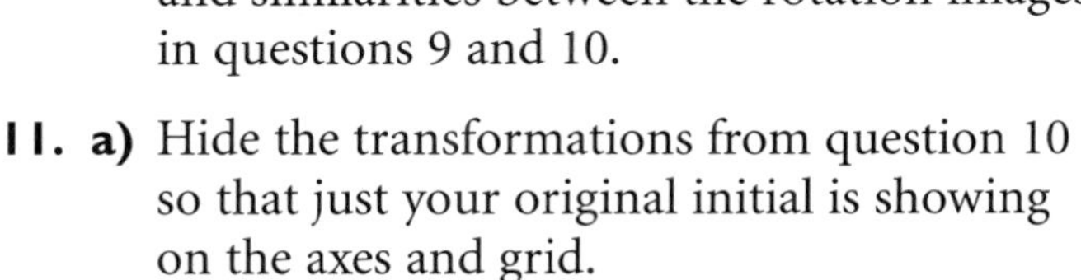

i) 45° **ii)** 60° **iii)** 90° **iv)** 180° **v)** 360°

9. **a)** Hide the last transformation so that just your original initial is showing on the axes and grid.
   b) Construct a point on your initial. Select the point and mark it as the centre.
   c) Select your initial. Transform it by rotating it 45°. Describe the effect of the rotation.
   d) Rotate the image from part b) 45°. Repeat until the image is at the same place as your original initial. Describe the effects of these rotations on the original initial. Save this file.

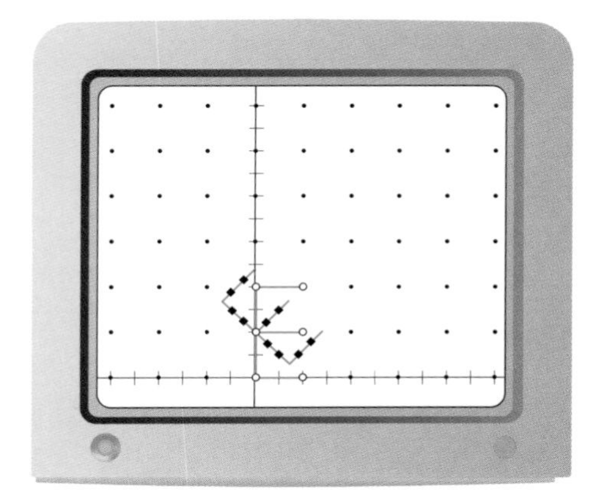

10. **a)** Hide the transformations from question 9 so that just your original initial is showing on the axes and grid.
    b) Repeat question 9, but construct a point outside your initial. Describe differences and similarities between the rotation images in questions 9 and 10.

11. **a)** Hide the transformations from question 10 so that just your original initial is showing on the axes and grid.
    b) Construct a point on your initial. Select the point and mark it as the centre.
    c) Select your initial. Transform it with a dilatation of scale factor 3. Describe the effect of the dilatation. Save this file.

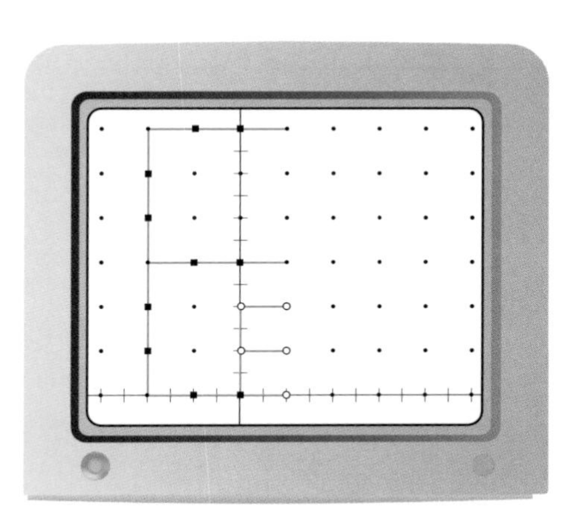

12. **a)** Hide the last transformation so that just your original initial is showing on the axes and grid.
    b) Construct a point, or use the same point as in question 11, on your initial. Select the point and mark it as the centre.
    c) Select your initial. Transform it with a dilatation of scale factor 0.5. Describe the effect of the dilatation. Describe differences and similarities between the dilatation images in questions 11 and 12. Save the file.

Practise

13. Describe each transformation.

a)

b)

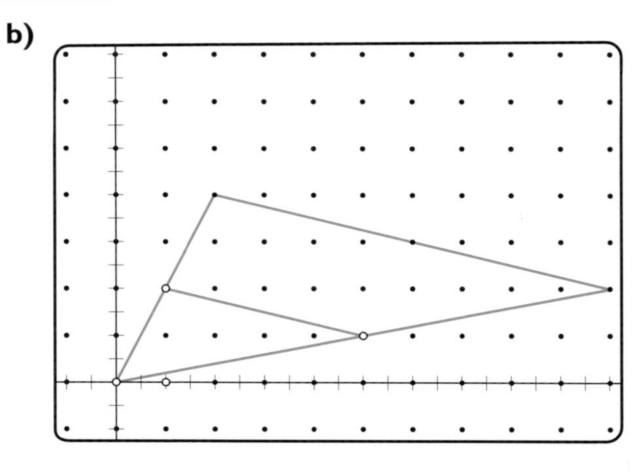

c)

d)

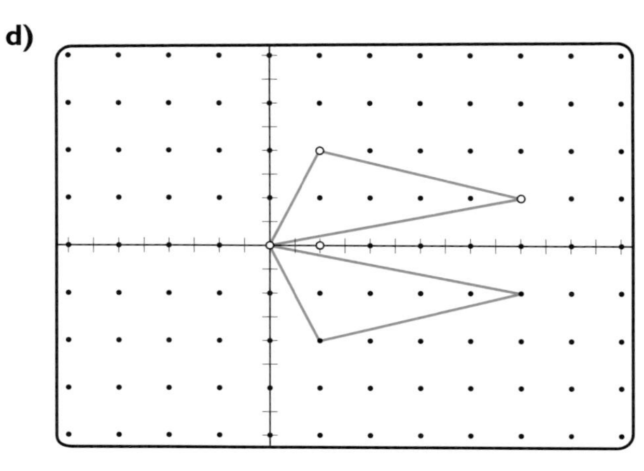

14. Recreate each transformation from question 13 using dynamic geometry software.

15. Construct a star or a parallelogram and perform the following transformations on it. Save each transformation in a separate file. Describe the effect of each transformation.

a) translation (5, −4)
b) translation (0, −5)
c) reflection in the horizontal axis
d) reflection in the vertical axis
e) rotation of 90° about the centre
f) rotation of 180° about the centre
g) dilatation about any point by a factor of 2
h) dilatation about any point by a factor of 0.5

10.3 Designing a Logo

Explore

What is the purpose of a logo? What characteristics should a logo feature? Explain.

Develop

A logo may be made up of text and graphic symbols. Text **font**, or type of a certain face and size, may be used to convey different images. A thick font often conveys strength and power. A script font conveys elegance. A slanted font conveys movement.

Colour, another aspect of a logo, also conveys different images. Navy blue, maroon, and dark teal are good choices for conveying a conservative image. Black and white produce a contemporary look.

1. You are going to start a business and you want to design an appealing and eye-catching logo that will symbolize your work. Your logo will appear on everything related to your business.
   a) Decide on a business. Come up with a single object, for example, an initial, name, shape, or graphic image that will be the basis of your business logo. Think in terms of a fairly simple, clean design.
   b) Sketch the object and identify transformations and symmetries you can use to design your logo using software.
   c) Using dynamic geometry software, design or drawing software, or dot paper, construct the object from part a).
   d) Design a logo based on the object constructed in part c) using the transformations and symmetries identified in part b). Remember that you can change the colour and style of various objects to make them more visually appealing. You may also fill the interior of some objects through shading.

Practise

2. Show your logo to two or three classmates, and examine your classmates' logos.
   a) List the transformations that you think were used in the construction. Describe any symmetry that you see.
   b) Give constructive suggestions to your classmates about their logos.
   c) Revise your logo based on the suggestions that you received from your classmates.

10.4 Career Focus: Sign Painter

Juliet is a sign painter. Her job involves laying out and printing letters, figures, and designs. Her signs appear on billboards, walls, cars, trucks, and buildings.

Sign painters must be accurate and precise with their hands, as well as detail-oriented. They must be able to work with a variety of materials, including paints, inks, coating materials, wood, fabric, and metal, and to operate power tools. Familiarity with design software is an asset.

Sign painters usually work for sign manufacturing companies with a team of skilled professionals, such as graphic designers. Sign painters with good business sense and communication skills are often self-employed.

1. What are the skills required to be a sign painter?
2. Why do you think it is important to work as an effective team player in this and other careers?
3. What does each of the following signs represent? What design features and transformations were used to convey the meaning of each sign?

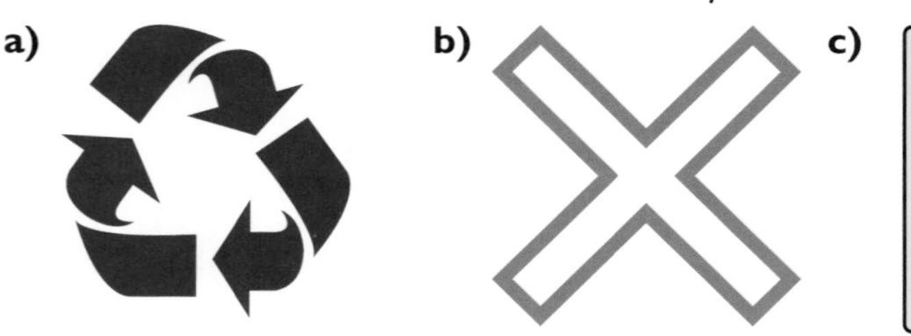

4. Examine signs in your school area or community. Find five that contain symbols or logos. Describe any transformations used in the symbols or logos. Describe the features of each sign that make it appropriate for the message it conveys.

10.5 Tiling a Plane

When tiling a floor or a wall, the most important aspect of the tiling job is that the surface gets covered. For example, tiling the walls of a shower with ceramic tiles is only effective if the ceramic tiles can be installed without overlapping or leaving any gaps.

Similarly, laying interlocking bricks on a patio or driveway is only effective if the bricks can be laid without overlapping or leaving any gaps.

Explore

Squares the same size, as shown above, can be used to **tile a plane**, or be placed edge-to-edge to cover a flat surface without spaces or overlap. The shapes below are other regular polygons. Each shape has equal sides and equal angles.

Determine which of these regular polygons can be used to tile a plane.

What are the characteristics of the shapes that can be used to tile a plane?

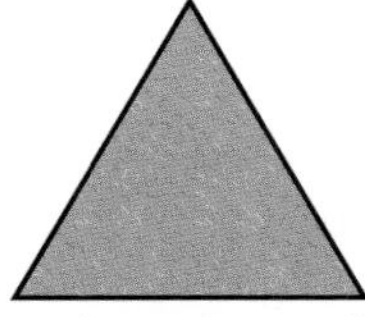

equilateral triangle

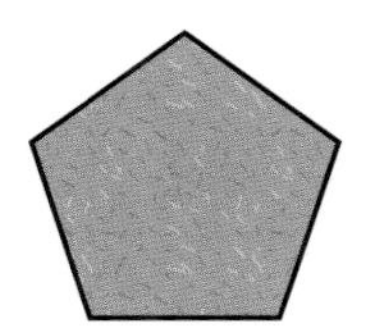

regular pentagon

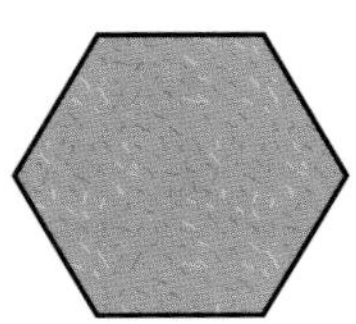

regular hexagon

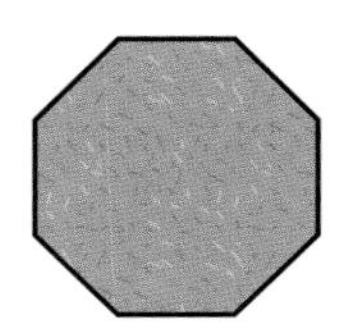

regular octagon

Develop

1. Determine which of these triangles can be used to tile a plane.

a) isosceles **b)** right **c)** scalene

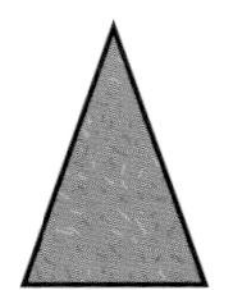
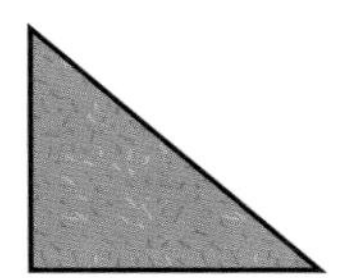
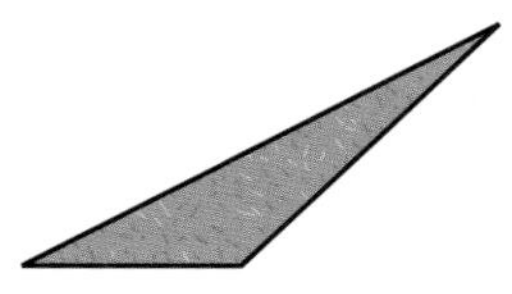

2. Which of these quadrilaterals can be used to tile a plane?

a) rectangle **b)** parallelogram **c)** rhombus

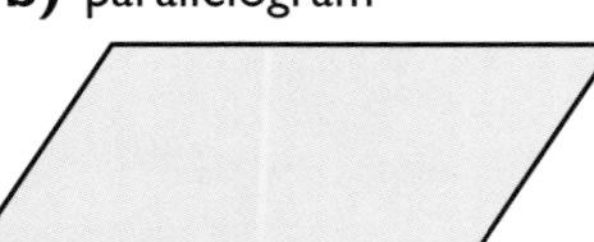

d) kite **e)** non-specific **f)** chevron

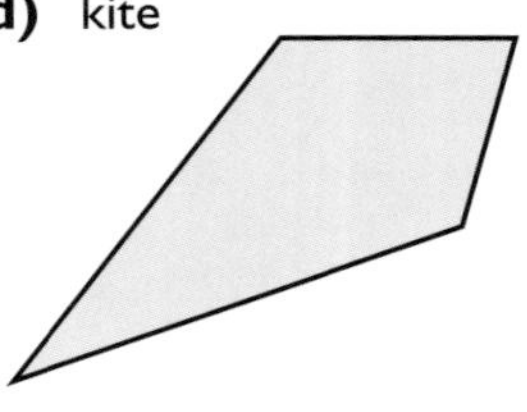
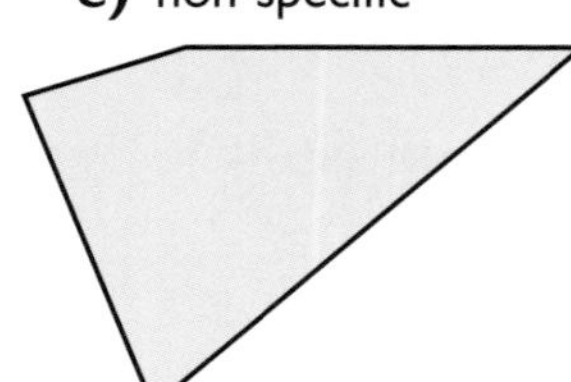
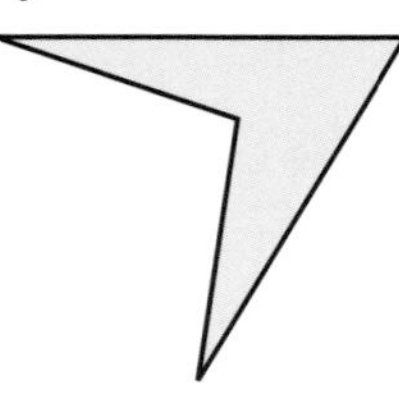

3. Two or more shapes can be used to tile a plane. Determine which of these pattern block shapes can be used in combination to tile a plane. What are the characteristics of the shapes that can be used to tile a plane?

Practise

4. Which of these shapes can be used to tile a plane?

a)

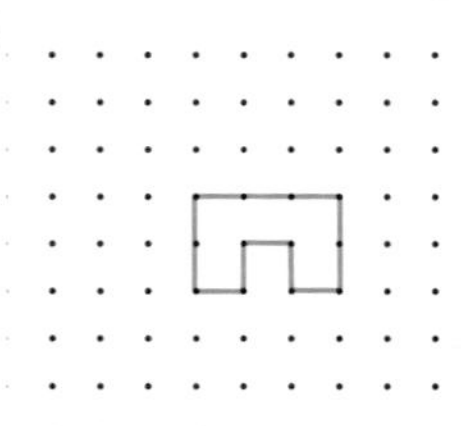

b)

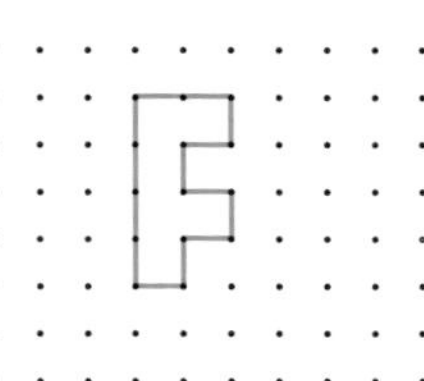

c)

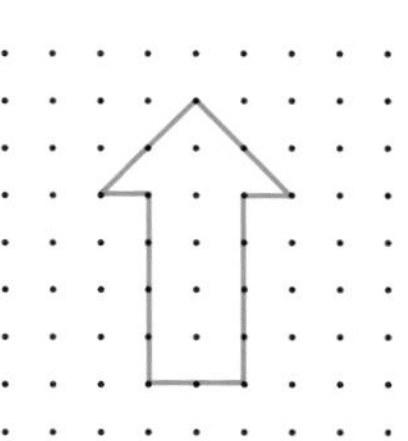

5. Determine if any combinations of shapes in question 4 can be used to tile a plane.

6. Explain which shapes and which combinations of shapes will tile a plane in terms of the angle measure where the vertices meet.

10.6 Designs Involving Tiling Patterns

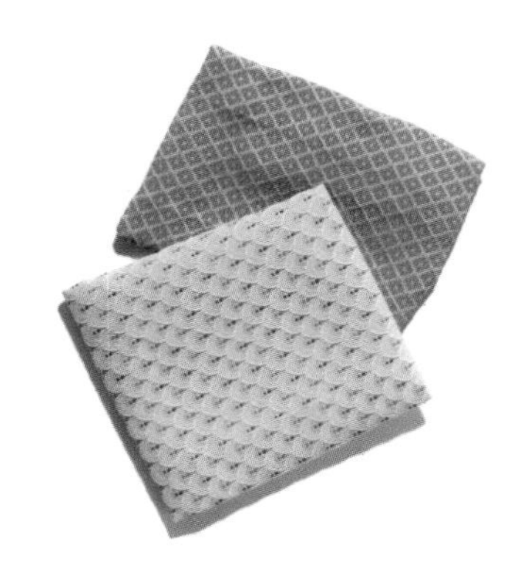

Wallpaper, fabric, gift wrap, and rugs are a few examples of art that often involve tiling patterns.

Explore

Use two or more shapes that you determined in Section 10.5 would tile a plane to create a design.

Develop

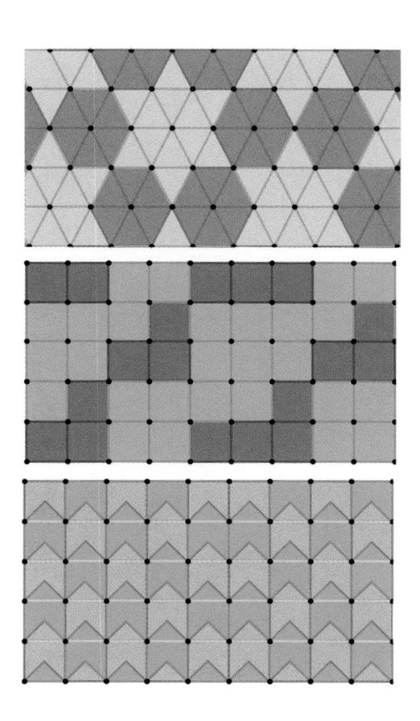

1. **a)** Tile a plane with equilateral triangles. Use colour with the triangles to create a different design from the one shown.
   b) Describe any transformations or symmetry in your design.
2. **a)** Tile a plane with squares. Use colour with the squares to create a different design from the one shown.
   b) Describe any transformations or symmetry in your design.
3. Start with a basic shape that can be used to tile a plane, modify it, and then create a design by tiling a plane with the modified shape. See the example at right.
4. Dutch artist M. C. Escher is renowned for his artwork involving tiling a surface with transformed shapes. Use the Internet to find out more about Escher and his art.

Access to Web sites about M. C. Escher and his art can be gained through the *Mathematics for Everyday Life 12* page of www.math.nelson.com.

10.7 Putting It All Together: Designing Geometrically

A You are starting a business that makes a product, such as wallpaper, fabric, or gift wrap, that involves tiling patterns.

Decide upon a business.

Create a logo.

- Describe key features of the business that should be reflected in its logo.
- Apply what you have learned to design a logo that reflects these key features.
- Describe the transformations and symmetry that you used.
- Describe the features that make the logo appropriate for the business.

Create a sample design of your product.

- Apply what you have learned about tiling patterns to create a design that involves a tiling pattern.
- Describe any transformations or symmetry in your design.

10.8 Chapter Review

1. Describe how symmetry and transformations have been used in each.

a)

b) South Korea

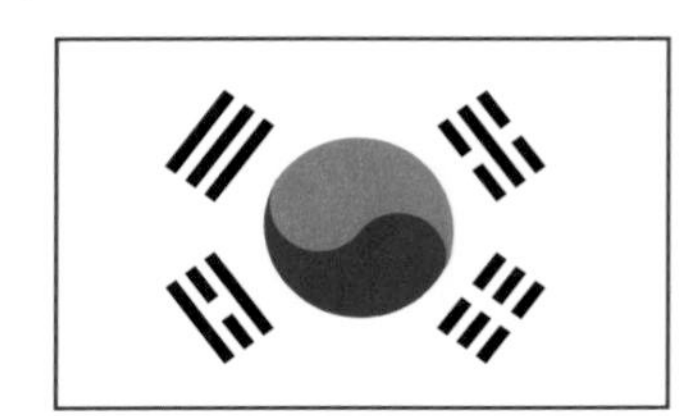

c)

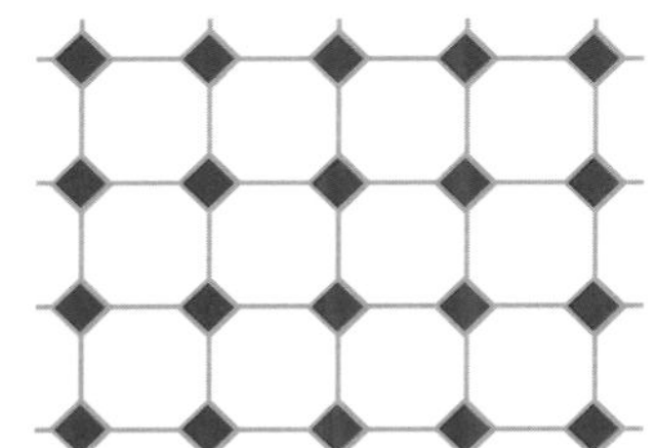

d)

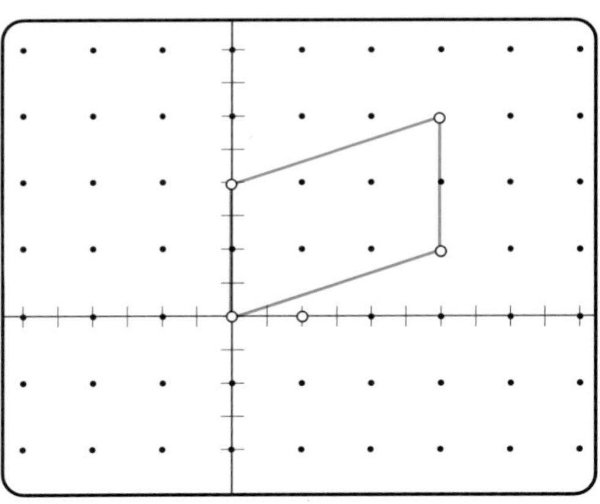

2. a) Using dynamic geometry software, construct a parallelogram with one point at the centre of the axes.

b) Reflect the parallelogram in the horizontal axis.

c) Rotate the image from part b) 90° about the origin.

d) Translate the image from part c) (–1, 3).

e) Dilatate the image from part d) by a factor of 2.

3. a) Think of a simple object on which to base a logo that would be suitable for a flower shop, a grocery store, or a jewellery store.

b) Using dynamic geometry software, design or drawing software, or dot paper, construct the object from part a).

c) Create the logo by transforming the object in part b).

d) Describe how you used transformations.

4. a) Identify three examples of shapes that can be used to tile a plane.

b) What are the characteristics of shapes that can be used to tile a plane?

5. a) Create a tiling pattern by transforming one shape.

b) Create a tiling pattern by transforming two or more shapes.

c) Describe any symmetry in your tiling patterns from parts a) and b).

Glossary

A

amortization period the time over which a mortgage or other loan is repaid

area the measure, in square units, of the surface of a two-dimensional shape

assignment of a lease an agreement that allows another person to take over an existing lease; the lease transfers to the new tenant.

B

bachelor apartment a suite in a building where the living room and bedroom are combined

bar graph (or **column graph**) a graph that compares data using equally spaced bars to represent the data

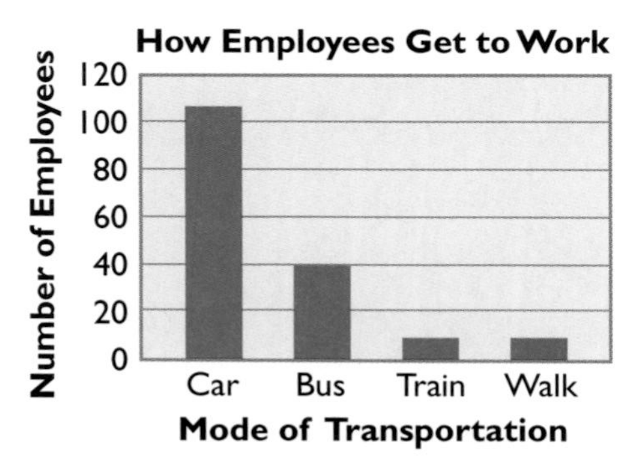

biased response a response favouring one side too much due to the way the survey questions were constructed

budget an organized plan for spending money

C

capacity the amount that a container will hold

circle graph (or **pie graph**) a graph that uses sectors of a circle to represent data as portions of a whole

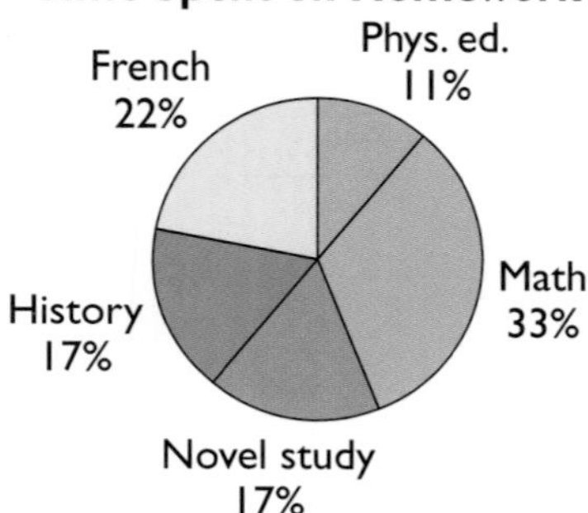

circumference the distance around a circle

column graph *See* bar graph.

condominium a type of property ownership, not a type or style of home; condominiums can be townhouses, low-rises, or high-rises.

condominium fee the amount paid by an owner of a condominium unit to cover maintenance and repair of common elements, such as the roof and hallways

congruent the same size and shape

cylinder a three-dimensional object with top and bottom faces that are congruent, parallel circles

D

data a set of facts, measurements, or values collected in a survey or experiment

dilatation a transformation in which a shape is enlarged or reduced about a point, resulting in an image that is similar, or proportional, to the original shape

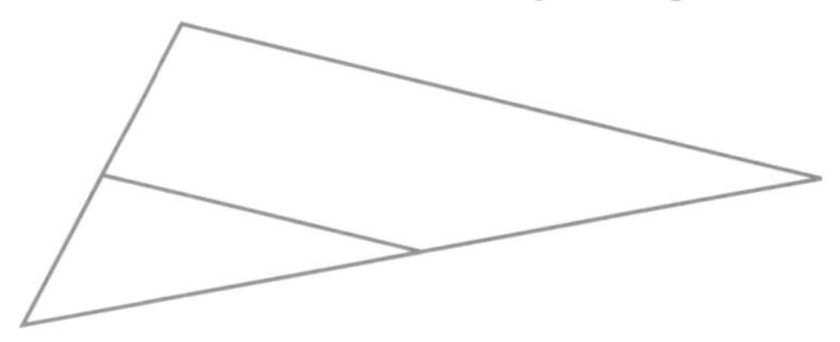

diorama a representation of a scene with three-dimensional models

double bar graph a graph that compares two similar sets of data using two sets of bars to represent the data

double line graph a graph that compares two similar sets of data using two sets of line segments

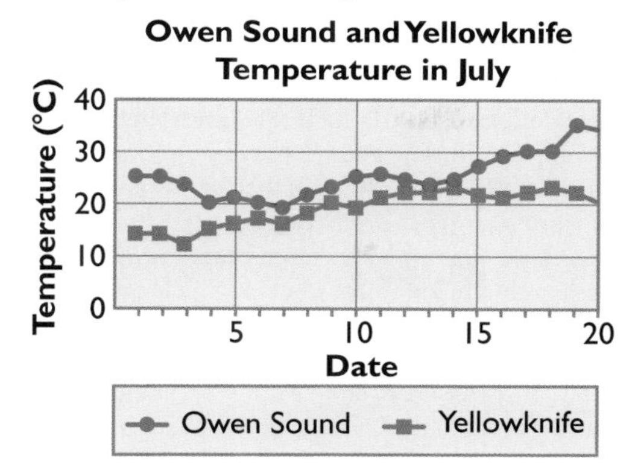

E

empirical probability *See* experimental probability.

enlargement a scale diagram or a scale model that is larger than the original where all lengths are proportional to those of the original

equal billing an equal amount paid each month for an otherwise variable expense such as heating, where any overpayment or underpayment is corrected at the end of the year

equally likely outcomes results in a probability situation having the same chance of occurring; for example, tossing heads and tossing tails are equally likely.

event a set of possible outcomes taken collectively; for example, the event of drawing an ace from a deck of cards has four outcomes: ace of clubs, ace of diamonds, ace of hearts, and ace of spades.

experimental probability the number of favourable outcomes divided by the number of possible outcomes determined by collecting data

F

favourable outcome the result being considered in a probability situation; for example, if tossing heads is being considered, then tossing heads is the favourable outcome.

fixed expense spending that stays the same from month to month, for example, car insurance payment

G

graph a diagram showing the relationship between variable quantities

H

hypotenuse the longest side of a right triangle

L

land transfer tax tax paid when real estate is purchased, based on the purchase price

lease a written tenancy agreement that gives the tenant a legal right to occupy a rental unit

liability insurance insurance that protects you from paying expenses related to damage or neglect for which you could be held responsible

line graph a graph that represents data using line segments; line graphs often show a change over time.

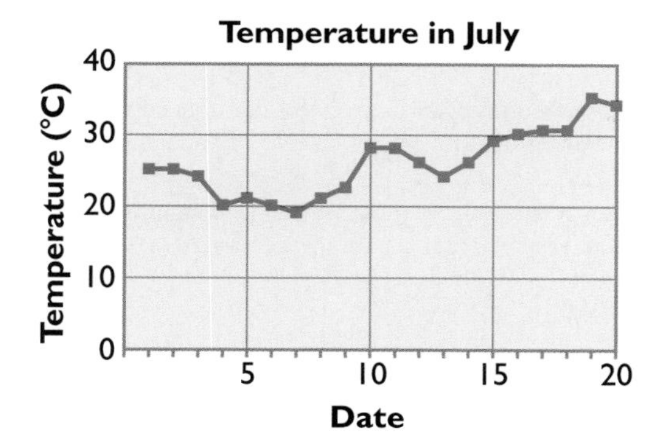

line (or **reflectional**) **symmetry** a property of a shape that fits onto itself after folding in a line

logo a distinctive symbol designed for and used by a company or organization to represent its image

M

mortgage a loan secured by real estate

O

outcome the result in a probability situation

P

perimeter the distance around an object

pictograph a graph that represents data using symbols

pie graph *See* circle graph.

PITH shelter costs; mortgage Principal, mortgage Interest, property Taxes, and Heating

population all the individuals or items under consideration

possible outcome a possible result in a probability situation; for example, the two possible outcomes of tossing a coin are heads and tails.

premium the amount charged by an insurance company for coverage

probability the likelihood or chance of something occurring; the number of favourable outcomes divided by the number of possible outcomes

property taxes taxes paid to a municipality (city, town, region) to pay for services such as garbage collection and recycling, parks and recreation; property taxes are a percent of the assessed value of the property.

Pythagorean theorem the relationship among the lengths of the sides of a right triangle: the square on the hypotenuse is equal to the sum of the squares on the other two sides.

R

radius the distance from the centre of a circle to the circumference

random numbers numbers generated without apparent pattern or reason

random sample a sample where every member of the population under consideration has an equal chance of being selected

rectangular prism a three-dimensional object with opposite faces that are congruent, parallel rectangles

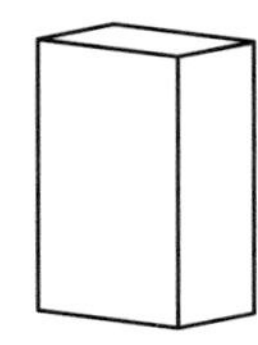

reflection a transformation in which a shape is flipped about a line, resulting in an image that is congruent to the original shape, but oriented the opposite way

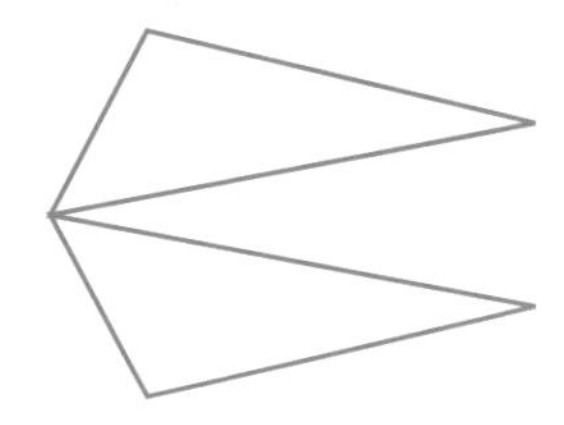

reflectional symmetry *See* line symmetry.

regular polygon a shape with all sides equal and all angles equal, for example, an equilateral triangle or a square

representative sample a sample typical of the characteristics of the population under consideration

right angle an angle with a measure of 90°

right triangle a triangle with one angle of 90°

risk factor a condition that is unfavourable for success

rotation a transformation in which a shape is turned about a point, resulting in an image that is congruent to the original shape, but oriented differently

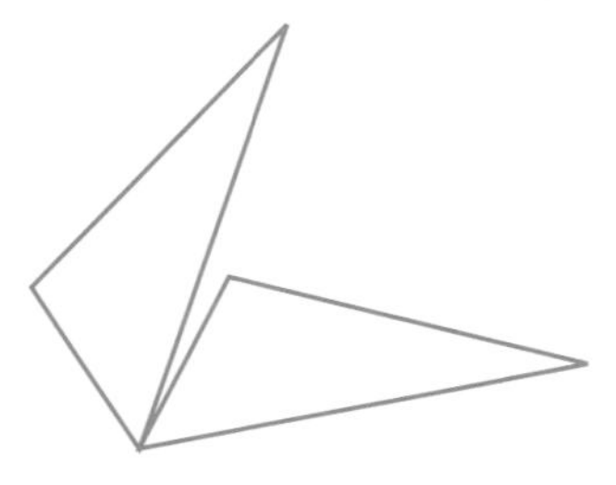

rotational symmetry *See* turn symmetry.

S

sample a small group selected from a population; data obtained from a sample is used to make predictions about the whole population.

sampling technique a process for collecting data from a sample, for example, conducting a survey, doing an experiment, counting, and measuring

scale drawing a drawing where all the lengths are proportional to those of the original

scale model a smaller or larger three-dimensional representation of an object where all the lengths are proportional to those of the original

shelter costs the cost of the mortgage principal, mortgage interest, property taxes, and heating for a home

simulation a probability experiment designed to model an actual event, for example, tossing a coin to determine whether the next child born in a family will be a boy or a girl

statistics the science of collecting, organizing, and interpreting data

Statistics Canada a government agency that collects, organizes, and analyzes data about many aspects of life in Canada

sublet renting of an apartment by a tenant to another person

surface area the total area of all outside surfaces of a three-dimensional object

survey a set of questions used to collect data; a survey may be by telephone, in person, or by mail.

symmetry a property of a shape that fits onto itself after a turn of less than 360° or after folding in a line

T

tally chart a chart used to record frequency

Number of Letters	Tally	Frequency
1	III	3
2	𝍸 𝍸 𝍸 IIII	19
3	𝍸 𝍸 𝍸 III	23
4	𝍸 𝍸 𝍸 𝍸	20

tenant a person renting an apartment

theoretical probability the number of favourable outcomes divided by the number of possible outcomes which can be calculated because the number of favourable outcomes and the number of possible are known facts

tile a plane place edge-to-edge to cover a flat surface without spaces or overlap

transformation a motion which creates an image of a shape: a translation, a reflection, a rotation, or a dilatation

translation a transformation in which a shape is slid in a straight line, resulting in an image that is congruent to and oriented the same way as the original shape

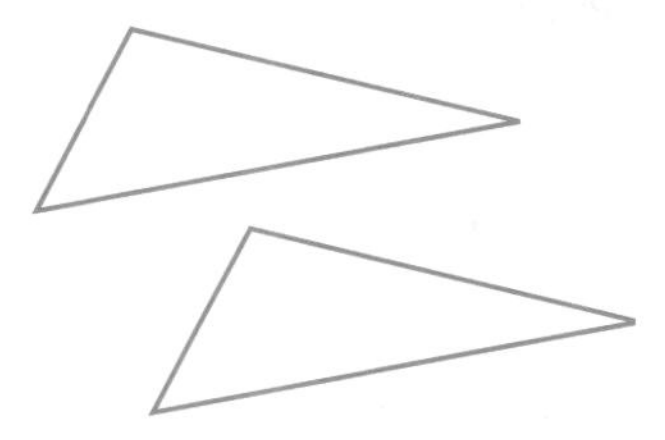

trial one action or a set of actions in a probability experiment; for example, in one experiment, one trial might be one roll of a die; in another, one trial might be tossing a coin four times.

turn (or **rotational**) **symmetry** a property of a shape that fits onto itself after a turn of less than 360°

U

utilities hydro, gas, and water

V

variable expense spending that changes from month to month, for example, groceries

volume the amount of space that an object occupies, measured in cubic units

Answers

Chapter 1: Data Graphs

1.1 Interpreting Graphs pp. 2–9

Develop

1. a) the average snowfall **b)** It is divided into increments of 50 cm. The scale allows for all the information to be seen. **c)** the cities **d)** Quebec City **e)** Vancouver **f)** Calgary and Toronto **g)** Answers will vary. For example, the bar graph allows the information to be assessed visually and is easier to read to get an impression, but exact values can be read from the table. **h)** Answers will vary. For example, advantages: the snow shovel is eye-catching and the replacement of the scale with the values given on the bars makes it possible to read exact values; disadvantages: without a vertical scale, it takes more time to interpret.

2. a) the percent of workers in each sector **b)** Answers will vary.

3. a) 99% **b)** 101% **c)** rounding **d)** 40–54 **e)** Answers may vary. For example, the graph above gives exact values. **f)** Since the number of years in each category is different, the categories have different weightings.

4. a) 10 seconds **b)** between 1960 and 1964; the line has the steepest decline between these points. **c)** between 1988 and 1992; the line goes up between these points. **d)** The data are continuous and well suited to a line graph. Line graphs are usually used to show change over time.

5. a) similar ups and downs for both lines **b)** The two statistics have more of an inverse relationship. **c)** No. When murders decrease, car crashes increase.

6. a) Police and fire **b)** Debt and Libraries, Parks and recreation **c)** Health, Social and children's services, and Other **d)** 100%, to complete a circle **e)** $870

Practise

8. a) double bar graph, compares two sets of discrete data **b)** line graph, shows the change of temperature over time **c)** pie graph, divides up the time spent on each subject as part of a whole **d)** double line graph, compares two sets of temperatures over many years **e)** bar graph, compares discrete data

9. a) They both represent the same data and are easy to read. The pie graph uses percents of total tax while the bar graph uses specific amounts spent on each area. **b)** Answers will vary. For example, the pie graph gives more information at a single glance.

10. a) 25% **b)** 25% **c)** 10% **d)** 66.7% **e)** 66.7% **f)** 60% **g)** 70% **h)** 25%

11. Some workers work in areas that are not identified on the graph, perhaps because the areas are too small or too specific.

1.2 Constructing Graphs pp. 10–13

Develop

1. a)

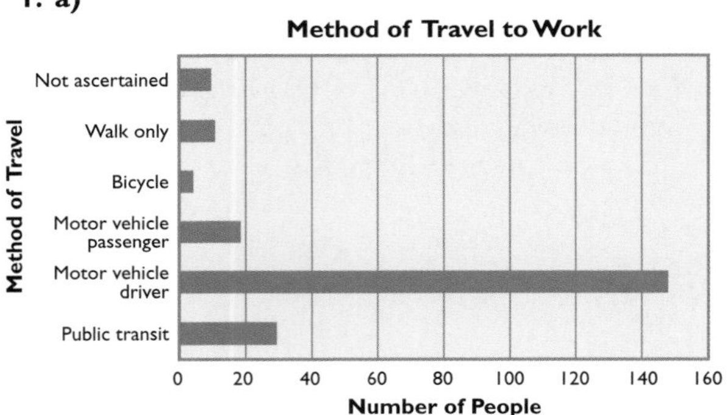

b) Motor vehicle driver **c)** Bicycle **d)** A bar graph displays discrete data well. **e)** They could point out the environmental impacts of so many people driving to work.

2. a)

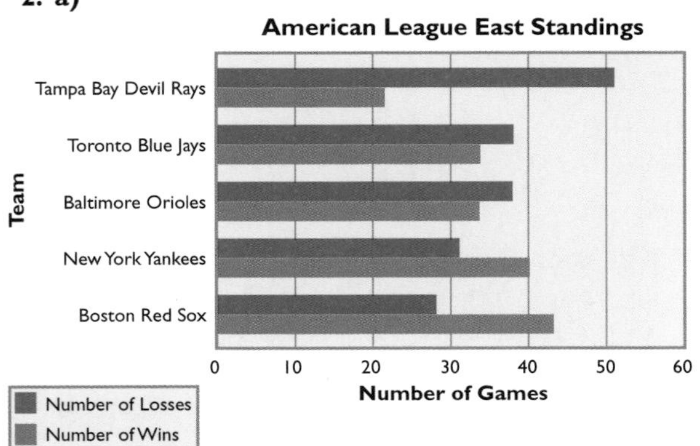

b) **i)** Boston Red Sox **ii)** Boston Red Sox **iii)** Tampa Bay Devil Rays **iv)** Tampa Bay Devil Rays **c)** The highest ranked team has the most wins and the least losses while the lowest ranked team has the fewest wins and the most losses. **d)** A double bar graph displays two sets of discrete data well.

3. a)

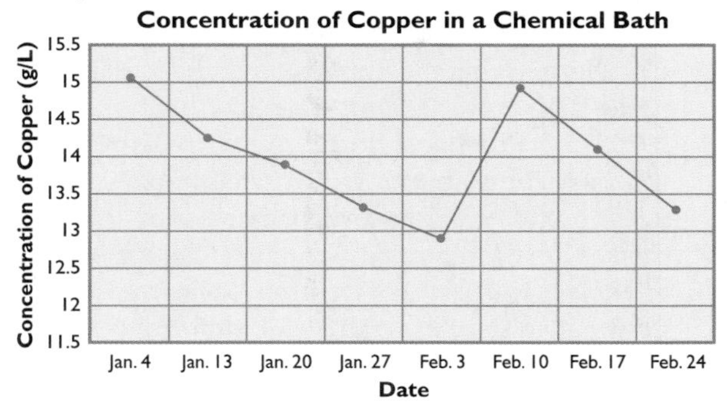

b) between Jan. 4 and Feb. 3, between Feb. 10 and Feb. 24 **c)** between Feb. 3 and Feb. 10 **d)** A line graph displays continuous data and change over time well. **e)** No, a line graph is the best.

Practise

4. a) 5°C **b)** 6°C **c)** 6°C **d)** 12°C **e)** 9°C **f)** 10°C

5. a)

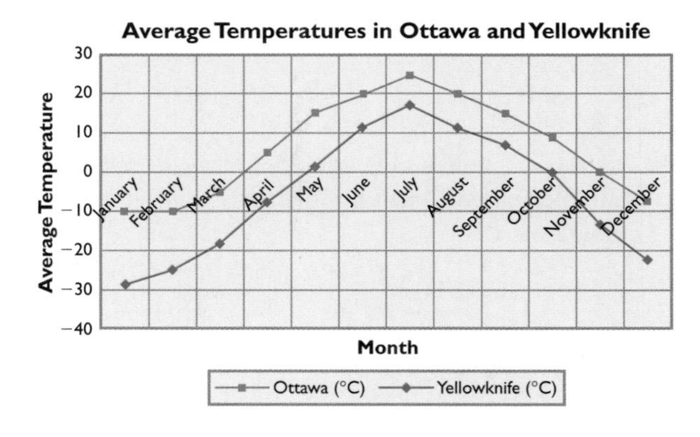

b) highest: Ottawa, July; lowest: Yellowknife, January **c)** greatest: January; least: July and September **d)** It is easier to use the graph because the difference is immediately visible. **e)** A double line graph displays two sets of continuous data and change over time well.

6. a)

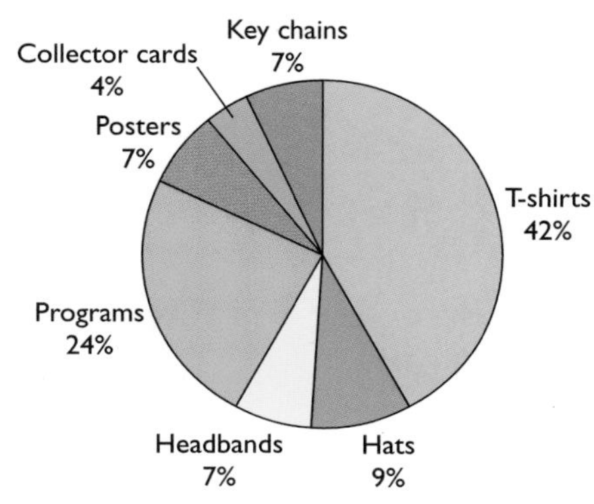

b) most: T-shirts; least: Collector cards **c)** It shows the various parts of merchandise production. **d)** You would need several pie graphs because one pie graph only shows one set of data.

7. a) Answers will vary. For example, a pie graph, to show parts of a whole **b)** Answers will vary. For example, a line graph, to show continuous data over time **c)** Answers will vary. For example, a triple bar graph, to compare three sets of discrete data **d)** Answers will vary. For example, a triple bar graph, to compare three sets of discrete data **e)** Answers will vary. For example, a double line graph, to compare two sets of continuous data

8. Answers will vary. For example, a line graph shows continuous data over time well.

1.3 Career Focus: Factory Worker pp. 14–15

1. a) to place blank boards into the machine after reading the order form to determine the number and type of boards to make **b)** He reads the order form to know how many of what items to make and how to make them. **c)** 50 **d)** 0.05 mm **e)** check boards for damage **f)** send them to the Inspection Department

2. to indicate that he was the person to complete the order

3. a) to know when to add copper to the chemical bath **b)** how to take measurements and plot graphs on his own

4. a) between Oct. 15 and Oct. 22, between Nov. 5 and Nov. 12, and between Nov. 19 and Nov. 26 **b)** between Oct. 8 and Oct. 15, between Oct. 22 and Nov. 5, between Nov. 12 and Nov. 19, and between Nov. 26 and Dec. 3 **c)** He added copper.

The graph increases after Oct. 15. **d)** He added copper to keep the level from dropping too low. **e)** They mark the range within which the concentration must be kept. **f)** It shows continuous data and change over time.

5. Answers will vary.

1.4 Constructing and Interpreting Graphs pp. 16–19

Develop

1. b) Graph types are bar or column, pie, and line. Answers will vary. For example, a bar or column graph displays data OK, but offers no sense of continuity; a pie graph does not suit the data; a line graph clearly shows changes in information, but exact values are not given. **c)** Answers will vary. For example, a line graph best conveys continuous data over time. **d)** between 2000 and 2050 **e)** Answers will vary. For example, medical advances and higher birth rates **f)** Answers will vary. For example, projecting from trends of the past

2. b) Answers will vary. For example, a double line graph compares two sets of continuous data over time. **c)** between 1950 and 2000 **d)** between 1950 and 2000 **e)** Answers will vary. For example, inflow of immigrants to United States and Canada from European countries **f)** Answers will vary. For example, reduced birth rates and emigration from Europe to other continents

Practise

3. b) Answers will vary. For example, a bar graph shows discrete data. **c)** Answers will vary. For example, a pie graph is not suitable because the data are not parts of a whole. **d)** Bowling **e)** Answers will vary. **f)** Answers will vary. For example, to lose or maintain weight

4. a) 25% **b)** purple and orange **c)** blue **d)** green

5. b) Answers will vary. For example, a double bar graph compares two sets of discrete data. **c)** Europe **d)** Answers will vary. For example, in both cities many people were born in Asia.

6. b) Answers will vary. For example, a pie graph shows the parts of a whole. **c)** Entertainment **d)** Savings **e)** Answers will vary. For example, spend less on entertainment

1.5 Misleading Graphs pp. 20–23

Develop

1. a) ii **b)** The vertical scale starts at $194 000 rather than at 0. **c)** 5.25% **d)** The change is evident and the break in the vertical scale draws attention to it.

Practise

2. a) The bottles change in two dimensions, height and width, giving the impression that, for example, production in the 2000s is nine times that of the 1970s rather than just three times. **b)** Answers will vary. For example, the A. Cola Company might use the graph to give the impression that its production has increased even more than it has. **c)** Answers will vary. For example, the graph could show bars or bottles which are a uniform width and change only in height.

3. a) 1st year, 2nd to 3rd quarter **b)** increasing more gradually **c)** The vertical scale starts at $120 000 rather than at 0. **d)** The vertical scale could start at 0 and show a break between 0 and 120 000 to draw attention to it.

4. a)

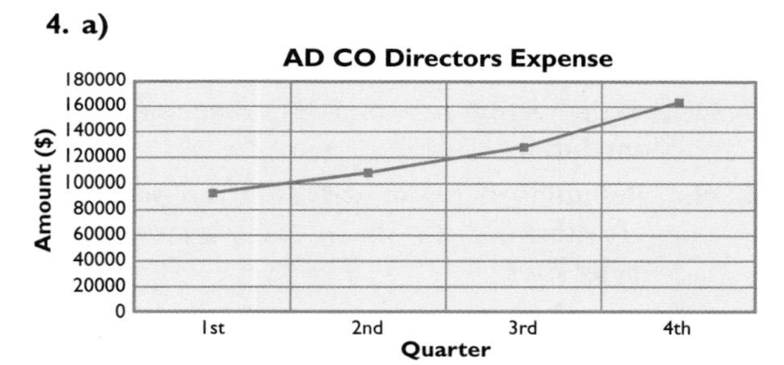

The vertical scale is compressed and the horizontal scale is expanded, creating the impression of very gradual increase.

b)

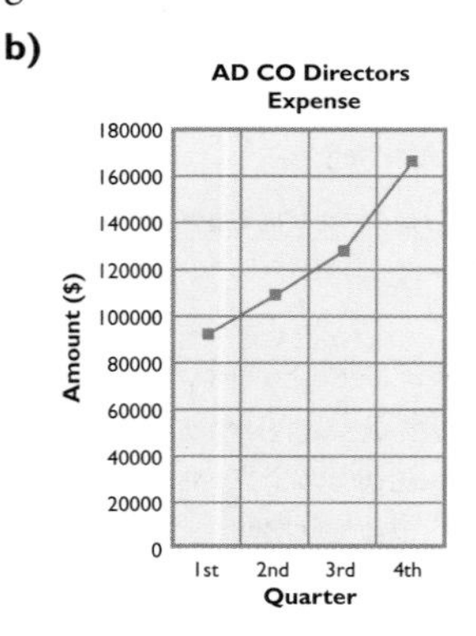

The vertical scale is expanded and the horizontal scale is compressed, creating the impression of a very rapid increase.

5. **a)** **i)** The cost of living is increasing quickly. **ii)** The cost of living is increasing very slowly. **b)** Start the vertical scale at 0 and show a break between 0 and 80 to draw attention to it.

6. Answers will vary.

1.6 Putting It All Together: Selling Shoes pp. 24–25

1. **b)** Answers will vary. For example, a line graph shows sales over several years well; a bar graph shows sales by size well; a pie graph shows all parts of September sales, assuming that there are only the four types of shoes.

2. decreased

3. Answers will vary. For example, they will be higher due to the general growth trend.

4. size 7

5. 74, assuming that she will sell the same proportion of size 9 shoes the following month

6. Women's Casual

7. Expand the vertical scale and/or compress the horizontal scale.

8. Her sales numbers reflect that she sells more women's shoes and that the most common sizes are smaller sizes.

1.7 Chapter Review pp. 26–27

1. Answers will vary.

2. **a)** Answers will vary. For example, it shows the parts of the whole budget. **b)** Education **c)** Public health **d)** Answers will vary. For example, any two of Education, Social services, and Police and fire departments

3. **b)** the groups that use the Internet more than 15 hours a week **c)** Answers will vary. For example, the percent of users who spend less time with family is equal to the percent of users who spend less time with friends, while for the other two groups, the percent of users who spend less time with friends is less than the percent of users who spend less time with family. **d)** A double bar graph compares two sets of discrete data well. **e)** Answers will vary. For example, a double line graph would not be suitable because the two sets of data are not continuous and do not show change over time.

4. Answers will vary. A line graph shows continuous data over time well.

5. **a)** between ages 1 and 2 **b)** Answers will vary. For example, you could continue the line in its upward trend.

6. **a)** Answers will vary. For example, it looks as if Geno has the largest share. **b)** Answers will vary. For example, yes. Abco is actually a larger section.

7. Answers will vary. For example, start the Sales scale at 0, but show a break between 0 and an amount just less than the first year's sales. Or, make the bar for the second year's sales twice as tall and twice as wide as the one for the first year's. The first suggestion is not misleading because the break draws attention to the scale; however, the second suggestion is misleading because the second bar gives the impression that sales are four times as great as those represented by the first bar.

Chapter 2: Collecting and Organizing Data

2.1 Sampling Techniques pp. 30–32

1. **a)** The sample is limited to two Canadian cities. It should include urban and rural Canadians from across the country. **b)** The sample is limited to communities in one fishing region. It should include urban and rural Canadians from across the country. **c)** The sample is limited to one shift on one day of the week. It should include a few cars from every shift. **d)** The sample is limited to customers at one restaurant. It should include people from around the city and away from any restaurants.

2. **a)** measuring **b)** survey **c)** survey **d)** measuring

3. **a)** Answers will vary. For example: Should more money be spent on security? Yes No No opinion **b)** not needed

Practise

4. **a)** survey; the data needed can be gathered by asking people questions. **b)** measuring; the date needed can be gathered by measuring the concentration. **c)** counting; the data needed can be gathered by counting vehicles passing through the intersection. **d)** experiment; the data needed can be gathered by testing some fireworks. **e)** survey; the data needed can be gathered by asking people questions.

5. a) Answers will vary. For example: Do you think that trains should be re-routed through less densely populated areas? Yes No No opinion **b)** not needed **c)** not needed **d)** Answers will vary. For example: It is fine for dolphins to be in marine zoos. Agree 5 4 3 2 1 Disagree

6. a) Answers will vary. For example, 500 adults of both genders from rural and urban areas selected from across the province **b)** Answers will vary. For example, survey by telephone, which allows for reaching people across the province from one location

7. Answers will vary. For example, experiments, and counting and measuring to ensure that the cereal is of adequate quality and quantity

8. a) Answers will vary. For example, counting and measuring **b)** Answers will vary. For example, in a line graph

9. a) Answers will vary. For example, biased: How would you rate the new hit TV series XYZ? Excellent 1 2 3 4 5 Poor; unbiased: How would you rate the new TV series XYZ? Excellent 1 2 3 4 5 Poor **b)** Answers will vary. For example, biased: Given the recent muggings in the area, do you believe the area has more than its share of crime? Yes No No opinion; unbiased: How would you rate the incidence of crime in the area? Heavy 5 4 3 2 1 Light

2.2 Statistics in the Media pp. 33–34

Develop

1. Answers will vary. Examples: What aspect of the environment is getting worse? Does the jobless rate include seasonal workers? self-employed workers? Why did the Ontario jobless rate rise?

2. a) Save the Children Canada **b)** the federal government **c)** the federal government **d)** Answers will vary. For example: What concerns you about growing up in Canada? **e)** The sample was a group of randomly chosen children aged 7 to 18 from 57 communities across Canada. The population is all children in Canada. **f)** 1200 children **g)** poverty **h)** any barriers to eating nutritious food, attending school, sleeping in a bed, and having clothes to wear **i)** Answers will vary. For example: Were the respondents asked to choose from a list of concerns or asked what their concerns were? **j)** Answers will vary. For example, well enough. The headline accurately identifies the top concern of the children surveyed.

3. Answers will vary.

4. Answers will vary.

2.3 Organizing and Interpreting Data pp. 35–38

Develop

1. a) The population is all secondary school students in Toronto. The sample is 35 secondary school students from Toronto. **b)** Answers will vary. For example, the bar graph **c)** most: Toronto; least: Calgary **d)** It depends on how the sample was chosen, but the sample is too small. **e)** Answers will vary. For example, a sporting goods store in Toronto could use the data to decide how many of the various teams' sweaters to stock.

Practise

2. a) Answers will vary. **b)** Answers will vary. For example: Which of the following prime time Tuesday shows do you watch most often? <list five shows> **c)** Answers will vary. **d)** Answers will vary.

3. b) Answers will vary. For example, City General. **c)** Answers will vary. For example, PVFD because it is the favourite **d)** Answers will vary. For example, television program creators

4. a) Answers will vary. **b)** Answers will vary. For example: Which of the following fast foods do you think is the most healthy to eat? <list five popular fast foods> **c)** Answers will vary. **d)** Answers will vary.

5. a) Answers will vary. **b)** Answers will vary. For example: Which of the following items do you wish you had more money to spend on? <list five items such as going to movies> **c)** Answers will vary. **d)** Answers will vary.

6. b) Answers will vary. For example, a bar graph shows discrete data well. **c)** most: 3 letters; least: 1, 8, and 9 letters **d)** Answers will vary. For example, from the table because the specific numbers are given **e)** Answers will vary. For example, a double bar graph

f) Answers will vary. For example, fewer long words in the children's book **g)** Answers will vary. For example, writers of promotional materials to determine the length of words to use depending upon their target audience

2.4 Career Focus: Telemarketer pp. 39–40

1. a) to sell things over the phone and to record his results **b)** to record information from sales calls **c)** He receives on-the-job training from his employer. **d)** These are the times when telemarketers are most likely to reach people at home.

2. Answers will vary.

3. a) Answers will vary. For example, randomly selected names from lists of cable subscribers from across the country **b)** Answers will vary. For example: Which of the following TV shows do you watch? <list the new series and the other shows of the same type> **c)** Answers will vary. For example, using a computerized tally **d)** Answers will vary. For example, a bar graph **e)** Answers will vary. For example, to get advertisers

4. a) Answers will vary. **b)** Answers will vary. **c)** Answers will vary. Examples: <list fast-food restaurants with a description of the food each serves> Rank each of the following fast-food restaurants from 1 Best value to 5 Worst value. <list restaurants> Rank each of the following fast-food restaurants from 1 Best-tasting food to 5 Worst-tasting food. <list restaurants> Rank each of the following fast-food restaurants from 1 Most healthy food to 5 Least healthy food. <list restaurants> **d)** Answers will vary **e)** Answers will vary. For example, a bar graph for each question **f)** Answers will vary. For example, any of the restaurants included to decide how to promote themselves if they ranked high or to decide how to improve themselves if they ranked low

2.5 Putting It All Together: Conducting a Survey p. 41

1–3. Answers will vary.

2.6 Chapter Review pp. 42–43

1. a) Answers will vary. For example, people in and around Barrie who grocery shop in Barrie **b)** Answers will vary. For example, 200 randomly selected residents of Barrie and the surrounding area

2. a) Answers will vary. For example, people who attend the community centre **b)** Answers will vary. For example, survey by personally interviewing people going into the community centre **c)** Answers will vary. For example: If a concession stand were to be opened at the community centre, which of the following types of food would you like to be able to buy at it? <list the foods that might be sold>

3. Answers will vary. For example, the sample is not suitable. It should include smokers and non-smokers from around the community, province, or country, depending on the population.

4. Answers will vary. Examples: How many Canadians were asked? What questions were asked? What does "not satisfied with" mean? Who conducted the survey? Who paid for the survey?

5. a) Answers will vary. For example, hockey is the activity that both groups spend the most time on; soccer is the activity that young adults spend the least time on; swimming is the activity that teenagers spend the least time on; and football is the activity that has the greatest difference in time spent between the two groups. **b)** Answers will vary. Examples: How was the sample selected? What question was asked? **c)** Answers will vary. For example, most suitable: the double bar graph compares two sets of discrete data well; least suitable: the double line graph because the data are not continuous

6. b)

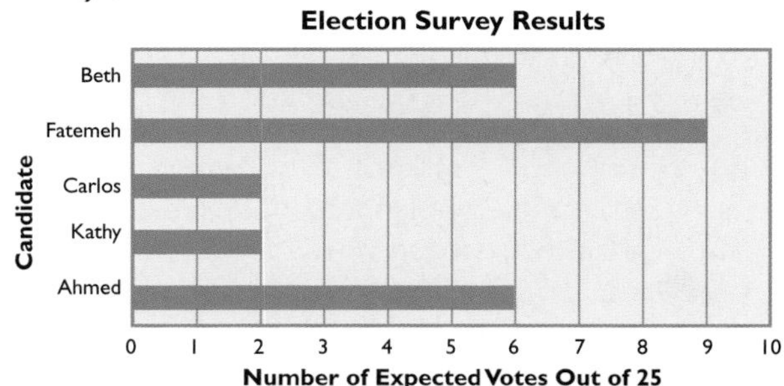

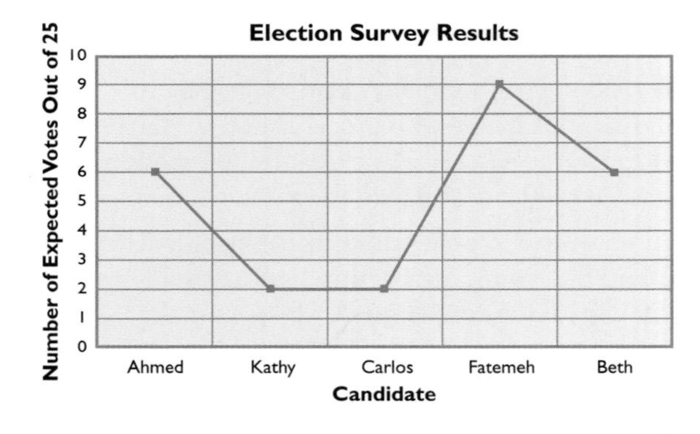

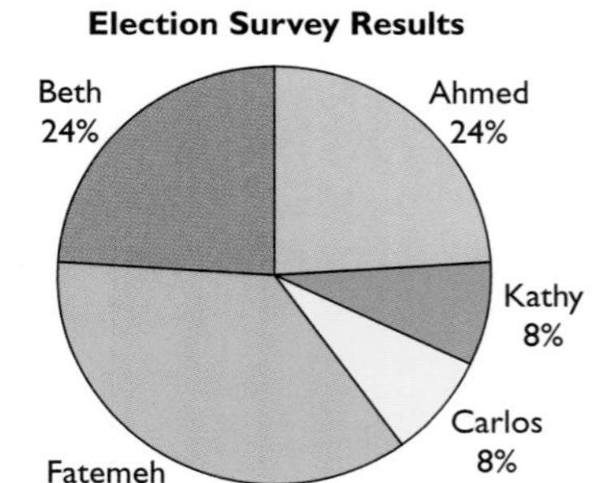

c) Answers will vary. For example, the bar graph because it displays discrete data well
d) Answers will vary. For example, Fatemeh has the most support; Beth and Ahmed are tied for second; Carlos and Kathy are tied for last with the least support.

Chapter 3: Probability

3.1 Making Predictions pp. 46–47

Develop

1. a) $\frac{1}{100}$ **b)** $\frac{5}{100}$ **c)** 0 **d)** 0 **e)** $\frac{13}{100}$ **f)** $\frac{79}{100}$
g) $\frac{8}{100}$ **h)** $\frac{92}{100}$ **i)** $\frac{13}{100}$ **j)** 1

2. a) no **b)** maybe **c)** April 25 **d)** Answers will vary. For example, to plan when they can go to their cottage by boat and when they can go over on the ice

3. a) the same **b)** Both are 0. Answers will vary. For example, Labour Day falling in July **c)** It is a certainty, 1. Answers will vary. For example, Canada Day falling in July **d)** April

Practise

4. 0.432, 0.410, 0.401, 0.373, 0.337, 0.291, 0.284

5. a) $\frac{18}{300}$ or $\frac{3}{50}$ **b)** $\frac{44}{300}$ or $\frac{11}{75}$
c) $\frac{192}{300}$ or $\frac{16}{25}$ **d)** $\frac{124}{300}$ or $\frac{31}{75}$

6. a) Answers will vary. For example, to better target their ads **b)** Answers will vary. For example, to better place their ads

3.2 Making Decisions pp. 48–49

Develop

1. a) 0.0005 **b)** risk: losing all her money; reward: winning fabulous prizes **c)** Answers will vary.

Practise

2. Answers will vary. For example, starting a fire and developing many diseases, such as all lung diseases, heart diseases, and stroke

3. a) i) 0.001 **ii)** 0.05 **iii)** 0.2 **iv)** 0.749 **v)** 1 **vi)** 0
b) Answers will vary. For example, risk: deciding to buy things you don't need and can't afford because you hope to get a good discount; reward: receiving discounts on purchases

4. a) Answers will vary. For example, surgery not being successful and surgery damaging your eyes
b) Answers will vary. For example, risks: damage to the eye or no improvement in vision; reward: corrected vision **c)** Answers will vary.

5. a) Answers will vary. For example, precipitation, high winds, extreme heat, and unseasonably cool temperatures on that day **b)** Answers will vary. For example, risks: less than perfect weather; rewards: the beauty of nature and pleasure of being outdoors **c)** Answers will vary.

6. a) Answers will vary. For example, his car being stolen, the security system failing, and a thief quickly damaging the car right after setting off the security system **b)** Answers will vary. For example, risk: car being stolen without a security system or the security system being faulty and not preventing a theft; reward: a feeling of security from having the security system **c)** Answers will vary.

7. a) Answers will vary. For example, Canada Savings Bonds earning so little interest that they won't even keep up with inflation, her friends' business having varying degrees of success or even failing, and the house becoming too expensive to maintain and having to be sold perhaps at a loss

b) Answers will vary. For example, risks: earning such a low rate of return on Canada Savings Bonds that it might not even keep up with inflation, losing some or all of her investment in her friends' business, or having to sell a house at a loss due to lack of money to maintain it; rewards: getting a guaranteed return on the Canada Savings Bonds, having the satisfaction of helping her friends, having a sound investment in her friends' business, and having the satisfaction of home ownership

c) Answers will vary.

3.3 Career Focus: Medical Office Receptionist pp. 50–51

1. greeting patients, finding files, checking OHIP cards, taking patients to examining rooms, answering the phone, making appointments, and answering general inquiries

2. Answers will vary. For example, interpersonal, organizational, and computer

3. Answers will vary. For example, kindness, patience, a calm manner, and flexibility

4. a) 30% **b)** protection against influenza and pneumonia **c)** mild side effects **d)** Answers will vary.

5. a) $\frac{1}{1\ 000\ 000}$ or 0.0001% **b)** $\frac{1}{1000}$ or 0.1%

c) risk: getting encephalitis; reward; gaining protection from measles, mumps, and rubella

d) Answers will vary. For example, getting measles without the vaccine

3.4 Comparing Probabilities pp. 52–53

Develop

1. a) favourable: the ice breaking on a specific date; possible: the ice breaking **b)** favourable: getting a card offering a specific discount; possible: getting a card

2. a) $\frac{1}{10}$ **b)** yes **c)** 0.1 and 10%

3. a) $\frac{1}{15}$, 0.066 666 …, or 6.67% **b)** $\frac{2}{18}$, 0.111 111, or 11.11% **c)** Answers will vary. For example, yes, with greater numbers, the probabilities would approach the theoretical.

Practise

4. a) $\frac{3}{25}$, 0.12, 12% **b)** $\frac{1}{250}$, 0.004, 0.4%

c) $\frac{1}{5000}$, 0.0002, 0.02%

5. Answers will vary.

6. 10 000 prizes

7. Steve Nash has higher accuracy on both. One out of two is greater than 43.6% and 88.3% is greater than 0.75.

3.5 Probability Experiments pp. 54–55

Develop

1. a) 2 **b)** 4, if order is important; 3, if order is not important **c)** 6 **d)** 52 **e)** 12

2. a) yes **b)** yes **c)** no

3. a) $\frac{1}{6}$ **b)–e)** Answers will vary.

Practise

4. a)–c) Answers will vary. **d)** $\frac{2}{8}$ or $\frac{1}{4}$ Answers will vary. **e)** Answers will vary.

5. a)–c) Answers will vary. **d)** $\frac{16}{36}$ or $\frac{4}{9}$ Answers will vary. **e)** Answers will vary.

3.6 Simulations pp. 56–58

Develop

1–2. Answers will vary.

Practise

3. a) two **b)** Answers will vary. For example, four coins where a tail represents a boy **c)–h)** Answers will vary.

4–6. Answers will vary.

7. a)–c) Answers will vary. **d)** months have different numbers of days

8. Answers will vary.

3.7 Putting It All Together: Spring Birthdays p. 59

Answers will vary.

3.8 Chapter Review pp. 60–61

1. Taylor, because her batting average is higher
2. **a)** 0.0001 **b)** 0.0001 **c)** 0.0005 **d)** Answers will vary. For example, risk: winning nothing; rewards: winning cash prizes and supporting a charity **e)** Answers will vary.
3. **a)** Answers will vary. For example, the GICs earning so little interest that they won't even keep up with inflation, the mutual funds not performing well and Cale losing some or all of his investment, and the blue chip stocks not performing well and Cale losing some or all of his investment **b)** Answers will vary. For example, risks: earning such a low rate of return on GICs that it might not even keep up with inflation, losing some or all of an investment in the mutual funds, or losing some or all of an investment in the blue chip stocks; rewards: getting a guaranteed return on the GICs, having a good return on the mutual funds, and having a good return on the blue chip stocks **c)** Answers will vary.
4. $\frac{17}{160}$, or 0.106 25
5. $\frac{2173}{7548}$, 0.28789, 28.789%
6. 381
7. 30 times, as the probability of getting heads is 50%
8. **a)–c)** Answers will vary. **d)** $\frac{1}{8}$ Answers will vary. **e)** Answers will vary.
9. **a)–b)** Answers will vary. **c)** four different stickers **d)–f)** Answers will vary.

Chapter 4: Renting an Apartment

4.1 Availability of Apartments pp. 64–66

Develop

1. Answers will vary.
2. **a)** A bachelor apartment does not have a separate bedroom. **b)** $593.33 **c)** fridge, stove, washer, and dryer **d)** apartments within houses **e)** utilities and phone service
3. Answers will vary. For example, location, number of bedrooms, availability, rent, what is included in the rent, a phone number to call

Practise

4. **a)** Answers will vary. For example, location downtown, availability **b)** Answers will vary. For example, number of bedrooms **c)** Answers will vary. For example, location in north-east side **d)** Answers will vary. For example, a phone number, the rent, what is included in the rent **e)** Answers will vary. For example, what is included in the rent **f)** Answers will vary. For example, location
5. Answers will vary.
6. Answers will vary.
7. Answers will vary.

4.2 Renting an Apartment pp. 67–70

Develop

1. **Quiz**

1. c
2. e
3. a
4. true
5. false
6. false
7. true
8. false
9. true
10. true
11. false
12. true

Practise

2. **a)** The landlord and tenant have a written agreement that specifies the rental for a period of 12 months. **b)** There is no tenancy agreement that specifies a fixed term. **c) i)** $9900 **ii)** $9880 **iii)** $10 400 **d)** Answers will vary. For example, location
3. **a)** $27.50 **b)** $27.20 **c)** $21.75 **d)** $18.90 **e)** $34.10 **f)** $30.34
4. **a)** $720.30 **b)** $883.15
5. **a)** $732.27 **b)** 5%, not legal
6. **a)** $800 **b)** $831.20 **c)** $31.20 **d)** $48
7. **a)** a payment equal to two months' rent, half of which is the first month's rent and half of which is the last month's **b)** to be protected against a tenant who vacates without notice **c)** $40.80 **d)** $699.72 **e)** $19.72

8. **a)** yes **b)** Omega **c)** Omega and Stephanie
9. **a)** no **b)** the landlord **c)** Trevor

4.3 Rights and Responsibilities of Landlords and Tenants pp.71–74

Develop

1.

Maintenance Responsibilities Quiz

1. true
2. true
3. true
4. false
5. true
6. true
7. true
8. true
9. true
10. false

Privacy Rights Quiz

1. A
2. B
3. C
4. B
5. C
6. A
7. C
8. C

Renewing and Terminating a Lease Quiz

1. false
2. true
3. true
4. false
5. true
6. true
7. false
8. false
9. true
10. true

Practise

2. a
3. b
4. September 30
5. June 30
6. March 26
7. **a)** Answers will vary. For examples, any five of the following: not paying the rent in full, often paying the rent late, conducting illegal activity, affecting the safety of others, disturbing the enjoyment of other tenants or the landlord, allowing too many people to live in the rental unit (overcrowding), or in subsidized housing not reporting income
b) causing undue damage, causing a serious allergic reaction in another person or the landlord, interfering with the normal enjoyment of the property by another tenant or the landlord such as causing undue noise, acting aggressively to others thus affecting their safety, or being a species or breed of animal that is inherently dangerous
c) Answers may vary. For example, any four of the following: landlord wants rental unit as own residence, or that of spouse or same-sex partner, or child or parent of one of them; landlord has agreed to sell the property to someone who wants all or part of the property for own residence, or that of spouse or same-sex partner, or child or parent of one of them; landlord plans major repairs or renovations that require a building permit and vacant possession; landlord plans to demolish the rental property; in a care home occupied for the sole reason of receiving therapy or rehabilitation, the rehabilitation or therapy program has ended; tenant of a care home needs more care than that available, or no longer needs the level of care provided

4.4 Career Focus: Building Superintendent p. 75

1. Answers will vary.
2. $3360
3. $117 280
4. Answers will vary. For example, learn maintenance skills, work with seniors

4.5 Monthly Apartment Costs pp. 76–78

Develop

1. a) \$19.44/month **b)** \$232.20/year **c)** \$211.68/year **d)** \$192.24/year **e)** \$18.14/month **f)** \$222.48/year

2. \$14.58

3. so that you have a record of everything that you own and can refer to it in the event of an insurance claim

4. a) The insurance company will pay up to \$1 000 000 if you are held liable and sued. **b)** \$300, because you pay less when an incident occurs

Practise

5. a) \$322.92 **b)** \$26.91

6. a) \$832 **b)** yes

7. a) \$972 **b)** \$2900

8. a) Insurance Total annual premium: \$192; PST: \$15.36; Total: \$207.36

Telephone Subtotal: \$52.75; GST: \$3.69; PST: \$4.22; Total current charges: \$60.66

Cable TV Subtotal: \$79.24; GST: \$5.55; PST: \$6.34; Total current charges: \$91.13

b) \$17.28

c) \$169.07

d) \$1500

4.6 Putting It All Together: Renting an Apartment p. 79

Answers will vary.

4.7 Chapter Review pp. 80–81

1. a) Answers will vary. For example: Is the pool indoors or outside? Is the apartment air conditioned? **b)** \$985 **c)** August 31 **d)** \$1800 **e)** \$926.10 **f)** no

2. Answers will vary. For example, insurance, phone, cable, heat, electricity, and water

3. in case someone gets hurts in your apartment and sues you

4. Answers will vary.

5. a) one week **b)** \$615.38 **c)** \$59 280

6. a) apartment assignment **b)** deductible **c)** termination date **d)** walk-up **e)** sublet an apartment **f)** contents insurance **g)** bachelor apartment **h)** utilities **i)** lease **j)** rental deposit **k)** eviction **l)** written notice **m)** Ontario Rental Housing Tribunal

7. a) \$585 **b)** telephone

Chapter 5: Buying a Home

5.1 Looking for a Home pp. 84–85

Develop

1. a) Answers will vary. **b)** Answers will vary.

2. Answers will vary.

3. a) Answers will vary. **b)** Answers will vary. **c)** Answers will vary.

Practise

4. a) Answers will vary. Examples given: **new** you get to select features, such as kitchen cabinets and floor coverings; everything is new and should work well. **resale** neighbourhood may be already established; home may be decorated and outdoors landscaped; since the home is already built, there should be no delays in moving in. **b)** Answers will vary. Examples given: **advantages** a single monthly fee covers several expenses that you would otherwise have to budget for separately; you own your home but don't have the cares and demands of upkeep. **disadvantages** you pay for some things that you might not want or need or might prefer to do yourself. **c)** detached: a home not attached to another; semi-detached: one of two homes attached on one side; duplex: one of two homes, where one is above the other; townhouse: one of several homes attached on the sides; mobile: no basement, so movable

5. Answers will vary. For example, some neighbourhoods have a reputation for a sense of community, safety, and proximity to desirable features, such as a lake, shopping, and entertainment. People will pay more to live in such neighbourhoods.

5.2 Buying a Home pp. 86–90

Develop

1. a) \$13 500 **b)** \$134 600 **c)** Yes, because the mortgage is for more than 75% of the value of the home.

2. **a)** \$9000 **b)** \$12 500 **c)** \$18 200 **d)** \$22 500 **e)** \$31 250 **f)** \$45 500

3. \$915

Practise

4. **a)** \$86 100 or \$105 900 depending on down payment **b)** \$26 500 **c)** No, because the mortgage is not more than 75% of the value of the home.

5. **all purchasers** land transfer tax; mortgage interest (assumes that homes are not paid for in cash); property insurance premiums; lawyer fees for reviewing the Offer to Purchase/Agreement of Purchase and Sale, drawing up mortgage documents, searching the title of the property, and tending to closing details; moving costs (assumes you have something to move); service hookup fees

   some purchasers GST, if a new home; a mortgage broker's fee, if a broker is used to find a lender; mortgage loan insurance premiums and fee, if mortgage loan is insured (insurance required if the mortgage is for more than 75% of the value of the home); property taxes paid in instalments added to mortgage payments (usually required if the mortgage is for more than 75% of the value of the home); an appraisal fee to have the value of the home assessed, if appraised (an appraisal is often required if the mortgage is not insured); a survey fee, if the lender requires a survey and you didn't ask the seller for one in the Offer of Purchase; adjustments to property taxes and utilities, if the seller has paid beyond the closing date; a home inspection fee, if they have the home inspected before signing the deal; repairs and renovations, if desired or needed; appliances, furniture, window coverings, other decorating, and tools, if needed; water quality and quantity certification fee, if the home has well service

6. a real estate agent to find the home; a mortgage broker to arrange the mortgage; an appraiser to determine the value of the home; a surveyor to survey the property; a lawyer to review the Offer of Purchase and the Agreement of Purchase and Sale, draw up mortgage documents, search the title of the property, and tend to closing details; and an inspector to inspect the home for structural faults

7. **a)** \$37 000 down payment, \$97 000 mortgage **b)** None, GST is not charged on resale homes. **c)** No, their mortgage was less than 75% of the value of the home. **d)** \$3177 **e)** \$40 177

8. \$134 000 is less than \$250 000, so use the first formula: \$134 000 × 0.01 = \$1340, \$1340 − \$275 = \$1065

9. **a)** \$1225 **b)** \$3423.50 **c)** \$5455 **d)** \$675 **e)** \$244 **f)** \$4460

10. Answers will vary.

11. \$168 130.84

12. **a)** \$160 000 **b)** GST is not charged on resale homes. **c)** Yes, their mortgage was more than 75% of the value of the home. **d)** \$1825 **e)** \$3888.86 **f)** \$53 888.86

13. Answers will vary. For example, what their income is, how much they have for a down payment, whether the home is new or a resale, if it's condominium ownership, and where the home is located

5.3 The Costs of Maintaining a Home pp. 91–93

Develop

1. **a)** \$2900 **b)** \$3560 **c)** \$4220 **d)** \$4700

2. **a)** Answers will vary. For example, older resale homes might require repairs sooner; new homes might have more low-maintenance features than older resale homes. **b)** Answers will vary. For example, no GST is paid on resale home; new homes should have few repairs for some time and more low-maintenance features.

3. **a)** A single monthly payment covers many expenses that other homeowners must budget for separately. **b)** Answers will vary.

4. **a)** No, she budgets for water and snow removal. **b)** Answers will vary. For example, heating, electricity, telephone, cable, home insurance, other maintenance and repairs

5. **a)** \$1440.29, \$120.02/month **b)** \$1746.22, \$145.52/month **c)** \$2436.81, \$203.07/month **d)** \$2724.48, \$227.04/month

6. **a)** when the mortgage is more than 75% of the value of the home **b)** A single monthly payment is convenient for the homeowner. The mortgage lender knows that the taxes are being paid when the payments are included with the mortgage payment.

7. **a)** taxes: \$223.07; heating: \$77.50; electricity: \$62.75; water: \$14.37; maintenance and repairs: \$300 **b)** \$1592.69 **c)** \$3355.41

8. a) taxes: $267.71; maintenance and repairs: $300 **b)** $1941.78 **c)** $4019.13

9. a) Answers will vary. For example, mortgage, property taxes, home insurance, maintenance and repairs, and maybe heating and electricity **b)** Answers will vary. Examples given: **owning a home** it is an investment, likely to go up in value; owners can renovate and decorate as they like within their budgets and enjoy having their own yards. **renting** fewer monthly expenses, no snow removal or yard maintenance responsibilities

10. Answers will vary.

5.4 Career Focus: Real Estate Agent pp. 94–95

1. $9120

2. a) the other company and its agent **b)** if she is both the buying and the selling agent **c)** $2280

3. a) $2400 **b)** $3600 **c)** $4800 **d)** $1389

4. $7725

5. $212 765.96

6. Answers may vary. Examples given: **advantage of selling privately** to save paying a real estate commission; **advantages of selling through a real estate agent** home exposed to many more potential buyers through MLS listing; agent has access to information to help set a selling price; agent shows your home; agent negotiates the deal.

5.5 Putting It All Together: Buying a Home p. 96

1. a) $166 900 **b)** Answers will vary. **c)** Estimates will vary. costs: land transfer tax; mortgage interest; property insurance premiums; lawyer fees; moving costs; service hookup fees; GST, if a new home; a mortgage broker's fee, if a broker is used to find a lender; mortgage loan insurance premiums and fee if insured (insurance required if the mortgage is for more than 75% of the value of the home); property taxes paid in instalments added to mortgage payments (usually required if the mortgage is for more than 75% of the value of the home); an appraisal fee to have the value of the home assessed, if appraised (an appraisal is often required if the mortgage is not insured); a survey fee, if the lender requires a survey and the buyer didn't ask the seller for one in the Offer of Purchase; adjustments to property taxes and utilities if the seller has paid beyond the closing date; a home inspection fee, if the buyer has the home inspected before signing the deal; repairs and renovations, if desired or needed; appliances, furniture, window coverings, other decorating, and tools, if needed; water quality and quantity certification fee, if the home has well service **d)** Estimates will vary. monthly costs: mortgage principal and interest, property taxes, heating, electricity, water, telephone, cable, home insurance, maintenance and repairs **e)** Answers will vary.

2. comparing amounts of money, estimating and calculating amounts of money, finding percents

5.6 Chapter Review p. 97

1. Answers will vary. For example, location, style, age, size, features, and construction

2. Answers will vary.

3. a) $61 300 **b)** $245 000

4. GST

5. a) $150 000 **b)** Yes, their mortgage was more than 75% of the value of the home. **c)** $1625 **d)** $3868.56 **e)** $43 868.56

6. a) Answers will vary. **b)** mortgage, property taxes, and heating **c)** mortgage principal, mortgage interest, property taxes, and heating **d)** PITH refers to mortgage Principal, mortgage Interest, property Taxes, and Heating. **e)** A single monthly payment covers many expenses that other homeowners must budget for separately.

Chapter 6: Household Budgets

6.1 Affordable Housing pp. 100–104

Practise

1. a) $400 **b)** $833.25 **c)** $800 **d)** $720 **e)** $308 **f)** $507.50

2. a) $512 **b)** $1066.56 **c)** $1109.33 **d)** $998.40 **e)** $394.24 **f)** $649.60

3. a) $33 600 **b)** $20/hour

4. a) $2205.65 **b)** $1689.48 **c)** $1443.98 **d)** $1306.11

5. **a)** $264 678 **b)** $304 106.40 **c)** $346 555.20 **d)** $391 833

   The monthly payments might be too large to handle over 10 years.
6. **a)** Answers will vary. **b)** Answers will vary. **c)** Answers will vary. **d)** Answers will vary.
7. **a)** no **b)** no **c)** no **d)** no
8. Answers will vary.

6.2 Components of a Household Budget pp. 105–7

Develop

1. Answers will vary.
2. **a)** Answers will vary. **b)** Answers will vary.
3. **a)** Answers will vary. **b)** Answers will vary. For example, use the average of the months.
4. **a)** seasonal changes **b)** to make budgeting easier **c)** Answers will vary. For example, they use past records, take into account price increases, and use the average of 12 months.

Practise

5. **a)** $76 **b)** $44.67 **c)** $47 **d)** $134 **e)** $25 **f)** $14
6. **a)** clothing and gifts **b)** Answers will vary. Examples given: **clothing** shops for clothes on an irregular basis; **gifts** birthdays and other events **c)** He could take the average amounts as the basis of his budget projections.
7. **a)** $358 **b)** with someone who pays the rent, such as his parents **c)** $698.67
8. **a)** phone: $36.67; entertainment: $64.67; clothing: $119; gifts: $22.67; miscellaneous: $18.33 **b)** $665.34
9. Answers will vary.
10. those containing special events, such as holidays, birthdays, and anniversaries
11. Impulse buying means buying something on the spur of the moment, without thinking about it before. A budget could help keep your mind focused on necessary items only.
12. **a)** $12.80/week **b)** $37.50/week
13. **a)** different budget categories **b)** different amounts needed **c)** different budget categories

6.3 Monthly Budget pp. 108–14

Develop

1. **a)** $15 **b)** gas and oil, clothing, personal items, medical and dental, presents, and subscription **c)** telephone and entertainment **d)** He is over by $17, but he is still all right because certain expenses, such as glasses and the subscription, will not reoccur next month. **e)** Answers will vary.

Practise

2. **a)** $475 **b)** $570 **c)** $95 **d)** Answers will vary. For example, decrease her clothing budget **e)** Answers will vary. For example, with a total monthly net income of $475, the total monthly expenses would be $475 if clothing was changed to $160 and personal items to $50.
3. **a)** $370 **b)** $1376 **c)** $225 **d)** over budget: telephone, transportation, clothing, entertainment, and other; under budget: food, personal items, and household items **e)** She is over budget. To verify, add all the actual bars and all the budget bars, and then compare.
4. **a)** clothing **b)** $100
5. $2281
6. **a)** $80 **b)** 12.5% **c)** clothing, entertainment, and personal items
7. **a)** $465 **b)** 59 hours
8. Answers will vary.
9.

Monthly net income	$3700
Monthly Expenses	
Fixed expenses	
savings	$370
rent	$900
public transportation	$90
Variable expenses	
clothing	$250
food	$400
entertainment	$100
personal items	$75
prescriptions	$40
household expenses	$85
Total expenses	$2310

6.4 Changing One Item in a Budget pp. 115–17

Practise

1.

Monthly Net Income	
Total	$1520

Answers will vary. Example given: Katie now has $1520 net income per month. She should increase her savings from $50 to $150. Since she is working full time, she may want or need to increase clothing from $150 to $200, and she may want to increase eating out from $60 to $90, and entertainment from $65 to $100. Changing her budget this way increases her total monthly expenses by $215, making them $635.

Moving in with her friend and not buying the car adds $350 for rent, including utilities, phone, cable, parking, about $200 for food, about $20 for long-distance calls, and about $25 for household items. Doing this would increase her total monthly expenses by $595 to $1230. She could do this.

Buying the car and staying at home adds $210 for car payments, $100 for insurance, about $120 for gas, and about $100 for maintenance and repairs, less $60 for her bus pass. Doing this would increase her total monthly expenses by $470 to $1105. She could do this.

Moving in with her friend *and* buying the car increases her total monthly expenses by $1065 to $1700. Doing this would put her over her monthly net income by $180. She could only do both by not increasing expenses associated with the full-time job.

2. Answers will vary. For example, change rent to $675.35 (an increase of about $26) and food to $334 (a decrease of $26).

3. Answers will vary. For example, Kyle has $456.20 less monthly net income. He could eat out less and reduce his food expense to $200 from $334. Since his workplace is just in a slow period, to reflect the lower income, he could reduce his savings to $180 from $290. He could reduce entertainment to $40 from $120 and clothing to $20 from $100. He could use his car less and reduce gas and oil expenses to $65 from $120. These changes would lower his total monthly expenses by $459, enough to work with the lower net income.

6.5 Career Focus: Furniture Refinisher p. 118

1. Answers will vary. For example, woodworking gives Rita experience in working with wood, math will help her with the accounting aspects of her business and with measuring materials, and English will benefit her when it comes to correspondence and communication in general.
2. **a)** $22 815 **b)** $438.75 **c)** $1578.04 **d)** $157.80/month
3. $24 765, 8.5%
4. $10 384.62 if renting, $14 400 if owning

6.6 Putting It All Together: Household Budgets p. 119

1–2. Answers will vary.

6.7 Chapter Review pp. 120–21

1. **a)** $600 **b)** $692.31 **c)** $2186.67
2. entertainment: $61; clothing: $72.67; gifts: $36.67; miscellaneous: $18
3. **a)** Answers will vary. For example, rent, insurance, and savings **b)** Answers will vary. For example, food, entertainment, and gifts
4. **a)** transportation and entertainment **b)** $240 **c)** $220 **d)** no
5. **a)** $32 **b)** **i)** food, **ii)** telephone, **iii)** transportation (all), **iv)** clothing, **v)** entertainment **c)** food, entertainment, and transportation (overall) **d)** telephone and clothing **e)** 8.3% **f)** $22 **g)** yes **h)** Answers will vary.

Chapter 7: Measuring and Estimating

7.1 The Metric System pp. 124–26

Develop

1. 6500 m
2. **a)** multiply by 100 **b)** multiply by 1000 **c)** multiply by 1000 **d)** multiply by 1000
3. 0.65 kg

Practise

4. **a)** centimetre **b)** metre **c)** kilometre **d)** millimetre **e)** gram **f)** kilogram **g)** gram **h)** millilitre **i)** litre **j)** litre
5. **a)** 380 **b)** 458 **c)** 2400 **d)** 4.35 **e)** 429 **f)** 380
6. **a)** 38.5 **b)** 4.32 **c)** 9.436 **d)** 3.25 **e)** 0.439 **f)** 0.0564
7. **a)** larger, 300 cm **b)** larger, 280 mm **c)** larger, 2400 m **d)** smaller, 48.5 cm **e)** smaller, 4.576 km **f)** smaller, 0.35 m **g)** larger, 2400 mm **h)** larger, 180 mm **i)** smaller, 2.495 m **j)** larger, 240 cm
8. **a)** 2.7 km **b)** 0.435 m **c)** 5 cm **d)** 4007 m **e)** 4300 mm **f)** 300 km **g)** 5 kg **h)** 0.67 kg **i)** 8300 mL **j)** 250 mL
9. **a)** multiply by 10 **b)** 64.8 cm long and 6.6 cm wide **c)** 7
10. **a)** divide by 10, to reduce the measurement to 1.8 cm **b)** 64
11. **a)** multiply by 1000, 12 000 m **b)** divide by 1000, 0.535 km
12. **a)** 1.96 m **b)** 153 cm
13. 24.55 km
14. **a)** larger, 5000 mL **b)** smaller, 2.5 L **c)** larger, 1360 mL **d)** smaller, 0.85 L **e)** larger, 7100 mL **f)** smaller, 5.525 L
15. **a)** 4000 mL **b)** 1500 mL **c)** multiply by 1000
16. **a)** 0.54 L **b)** 0.75 L **c)** divide by 1000
17. $12.13

7.2 Measuring Lengths pp. 127–29

Practise

1. **a)** inches **b)** inches **c)** feet **d)** inches **e)** miles
2. Answers will vary. Examples given:
   a) $\frac{3}{4}$ inch **b)** 8 inches **c)** $\frac{1}{2}$ inch
   d) 5 feet 5 inches (or 65 inches) **e)** 5 feet (or 60 inches) **f)** 3 feet 5 inches (or 41 inches) **g)** 4 feet 8 inches (or 56 inches)
3. Answers will vary. Examples given: **a)** 2 cm **b)** 20 cm **c)** 1.4 cm **d)** 1.65 m (or 165 cm) **e)** 1.53 m (or 153 cm) **f)** 1.04 m (or 104 cm) **g)** 1.42 m (or 142 cm)
4. Answers will vary. Examples given: **a)** 5.7 cm, 57 mm **b)** 3.2 cm, 32 mm **c)** 2.7 cm, 27 mm **d)** 12.5 cm, 125 mm
5. Both sheets are 4 feet wide. One is 4 feet long and the other is 8 feet long.
6. **a)** The first measurement is the thickness of a deck board, 2 inches or $\frac{5}{4}$ inch. The second is the width of each deck board, 6 inches. The third is the length of a deck board, 8 feet, 10 feet, 12 feet, or 16 feet.
   b) 40 boards of 2" × 6" × 16' or $\frac{5}{4}$" × 6" × 16'
7. **a)** Answers will vary. For example, screws, lumber, bolts, nails, and pipes **b)** Answers will vary. For example, paint, glue, rope, masking tape, and pails
8. Yes, all of the van's measurements are less than the dimensions of the garage.
9. $198.14

7.3 Estimating Distances pp. 130–31

Develop

1. Estimates will vary. Examples given: **a)** 18 cm **b)** 10 cm **c)** 15 cm **d)** 60 cm **e)** 14 m **f)** 40 cm
2. Estimates will vary. Examples given: **a)** 7 inches **b)** 4 inches **c)** 6 inches **d)** 2 feet **e)** 15 yards **f)** 16 inches
3. **a)** 12 mm **b)** 5 m **c)** 550 m **d)** 2 m
4. Estimates will vary. Examples given: **a)** 60 cm **b)** 30 cm **c)** 12 cm **d)** 1 m **e)** 2.5 m
5. **a)** 2 feet **b)** 3 feet **c)** 25 feet **d)** 8 inches
6. Estimates will vary. Examples given: **a)** 7 feet **b)** 1 foot **c)** 18 inches **d)** 150 yards **e)** 5 yards
7. Answers will vary.

7.4 Estimating Capacities pp. 132–33

Develop

1. **a)** millilitre **b)** litre **c)** millilitre **d)** litre
2. Answers will vary. Examples given: **a)** egg shell **b)** pop can **c)** pop bottle **d)** large juice can **e)** water cooler jug **f)** kiddie pool

Practise

3. **a)** 50 L **b)** 75 mL **c)** 7 L **d)** 350 mL **e)** 200 mL **f)** 40 L **g)** 2 L **h)** 5 L **i)** 625 mL
4. Estimates will vary. Examples given: **a)** 300 mL **b)** 10 L **c)** 50 000 L **d)** 80 L **e)** 15 mL **f)** 350 L
5. 4.8 L

6. 1.5 L
7. 200
8. 1.345 L
9. Answers will vary. Examples given: **a)** 950 mL, should be enough for two coats **b)** 3.7 L, need at least this much **c)** 284 mL, more than enough **d)** 3.7 L, should be enough for two coats

7.5 Estimating Large Numbers pp. 134–35

Practise

1. **a)** Answers will vary. For example, 350; about 11 rows with about 32 cars in a row **b)** Answers will vary. For example, 525; about half the cars with one person and half with two
2. Answers will vary. For example, 80; about 10 down and 8 across
3. **a)** Answers will vary. For example, 528; 11 floors, about 22 windows per floor on long side, 2 per floor on short side **b)** Answers will vary. For example, 308; 11 floors, 7 balconies per floor per side, 1 balcony per apartment, average of 2 people per apartment
4. **a)** Answers will vary. For example, count the number of shelves and count the number of books on half a shelf. **b)** Answers will vary. For example, count the number of rows of shelves, count the number of sections of shelving in a row, and count the number of books in half a section of shelving.
5. Answers will vary. Examples given: In 1 a), it was assumed the rows were full and there were no part rows. These assumptions could balance each other. In 1 b), it was assumed that half the cars had one person and half had two, that no one arrived at the mall by other means, and that what is shown is the entire parking lot. These assumptions would make the answer inaccurate.

   In 2, it was assumed that the quarters are to be lying flat, touching, but not overlapping. These assumptions, if true, lead to obtaining a fairly accurate answer.

   In 3 a), it was assumed that all floors are the same, even the ones hidden behind trees, and that each balcony has two windows. The first assumption is reasonable and shouldn't affect accuracy. The second one, if not correct, would lead to inaccuracy. In 3 b), it was assumed that there was one balcony per apartment and an average of 2 people per apartment. The first assumption is reasonable and shouldn't affect accuracy. The second one, if not correct, would lead to inaccuracy.

7.6 Career Focus: Decorating Store Clerk p. 136

1. **a)** 3 blinds at $51\frac{7}{8}$ inches × $39\frac{3}{4}$ inches; 1 blind at $79\frac{7}{8}$ inches × $70\frac{1}{2}$ inches; 3 blinds at $56\frac{1}{8}$ inches × $42\frac{5}{8}$ inches; 1 blind at $52\frac{5}{8}$ × $36\frac{3}{16}$ inches

   b) 3 blinds at $51\frac{7}{8}$ inches × $39\frac{3}{4}$ inches cost $180 each; 1 blind at $79\frac{7}{8}$ inches × $70\frac{1}{2}$ inches costs $312; 3 blinds at $56\frac{1}{8}$ inches × $42\frac{5}{8}$ inches costs $180 each; 1 blind at $52\frac{5}{8}$ × $36\frac{3}{16}$ inches costs $180.

   c) $1807.80 **d)** $723.12; $1084.68

7.7 Putting It All Together: Estimating p. 137

1–2. Answers will vary.

7.8 Chapter Review pp. 138–39

1. **a)** centimetre **b)** metre **c)** kilometre **d)** centimetres or metres **e)** millimetres
2. **a)** 200 cm **b)** 480 mm **c)** 3200 m **d)** 39.2 cm **e)** 3.45 km **f)** 0.48 m
3. **a)** $\frac{3}{4}$ inch, 18 mm

   b) $\frac{7}{8}$ inch, 21 mm

   c) $\frac{7}{16}$ inch, 17 mm

   d) $\frac{15}{16}$ inch, 23 mm

   e) $1\frac{1}{16}$ inches, 26 mm

   f) $1\frac{1}{8}$ inches, 28 mm
4. Answers will vary.
5. **a)** 4.8 km **b)** 15.6 km
6. **a)** 16 cm **b)** 3 cm **c)** 50 cm **d)** 8 m

7. **a)** inches **b)** feet and inches **c)** inches **d)** feet **e)** miles
8. **a)** $36\frac{1}{2}$ inches **b)** 7 feet **c)** 9 yards
9. Estimates will vary. Examples given: **a)** 15 feet, 5 m **b)** 6 feet, 2 m **c)** 10 feet, 3 m
10. **a)** 13 L **b)** 250 mL **c)** 15 L
11. **a)** 3000 mL **b)** 1250 mL **c)** 0.75 L
12. Answers will vary. For example, there could be 10 752 stitches. The 3 cm by 3 cm actual size square has 17×17, or 289, squares; therefore, one square centimetre would have $289 \div 9$, or about 32, stitches. The 21 cm by 16 cm picture would have about $32 \times 21 \times 16$, or 10 752, stitches.
13. **a)** They are both used in Canada. **b)** It is based on a single multiple, 10.

Chapter 8: Measurement and 2-D Design

8.1 The Pythagorean Theorem pp. 142–43

Develop

1. **b)** Answers will vary. For example, use a square corner. **f)** The sum of the areas of the two smaller squares equals the area of the largest square.

Practise

2. **a)** 468 **b)** 277 **c)** 808 **d)** 81.09 **e)** 677.69 **f)** 193.96
3. **a)** 6.7 **b)** 8.9 **c)** 11 **d)** 14.1 **e)** 15.8
4. **a)** 0.1 m **b)** 0.0 m **c)** 2.8 m
5. 70.7 cm
6. **a)** 30 feet **b)** No, or the diagonal would have measured 30 feet.
7. 60 feet
8. Answers will vary. For example, he could measure the diagonals which should be about 12.1 m each.

8.2 Calculating Perimeter and Area pp. 144–47

Practise

1. **a)** iii **b)** iv **c)** ii **d)** i
2. **a)** 12.25 m^2 **b)** 8.82 m^2 **c)** 7.1 m^2 **d)** 26 m^2
3. **a)** 14.4 m **b)** 17.2 m **c)** 11.3 m **d)** 28 m
4. 11.9 m^2
5. to put caution tape around the asphalt-laying machine
6. **a)** 300 m^2 **b)** 6
7. **a)** 108 square feet **b)** 10 **c)** 42 feet **d)** 5
8. **a)** Calculate the total area of the property and subtract the area of the house and the area of the driveway. **b)** 1790 square feet **c)** 199
9. **a)** 2000 m. The length required is 2 km. **b)** $69 000

8.3 Estimating Perimeter and Area pp. 148–49

Develop

1. Answers will vary.

Practise

2. Answers will vary.
3. **a)** Answers will vary. For example, 4 m^2
   b) Answers will vary. For example, 40 m^2
   c) Answers will vary. For example, 40 m^2 + 10 m^2, or 50 m^2 leaving space for waiters/waitresses and customers to pass
4. Answers will vary.
5. Answers will vary.
6. Answers will vary.
7. Answers will vary.
8. Answers will vary. For example, 60 m^2
9. Answers will vary.

8.4 Career Focus: Flooring Installer p. 150

1. **a)** 2276 square feet **b)** 200 **c)** $8346
2. **a)** 72 feet **b)** 18 **c)** 300 square feet were needed, but 49 square feet were left over.

8.5 Enlargements pp. 151–52

Develop

1. original: 35 square inches; enlargement: 140 square inches. The area increases four times when the dimensions are doubled.

Practise

2. **a)** The new screen will have an area four times as large as the original. **b)** old: 99 square inches, new: 396 square inches **c)** ii

3. **a)** 8 m by 20 m **b)** 16 times
4. **a)** 3 times as long and wide **b)** 9 times **c)** original: 56 cm^2; enlarged: 504 cm^2
5. 6.25 times
6. 36 times
7. Use an example. original 10" × 12", area = 120 square inches; 3 × 10 = 30 and 3 × 12 = 36; enlarged 30" × 36", area = 1080 square inches; 1080 ÷ 120 = 9
8. Double both measurements.
9. **a) i)** equal **ii)** The original is four times as large as the new circle. **iii)** The new circle is four times as large as the original circle. **b)** original = 0.1256 m^2 **i)** 0.1256 m^2 **ii)** 0.0314 m^2 **iii)** 0.5024 m^2
10. **a)** four times as large **b)** original: 112.5 cm^2; enlarged: 450 cm^2

8.6 Scale Drawings pp. 153–54

Develop

1. **e)** Answers will vary. For example, 18 in 6 rows of 3 units placed end to end, parallel to the side shelves
2. **a)** 4.5 cm **b)** 0.5 cm or 5 mm **c)** 0.5 cm **d)** 3 cm **e)** 2 cm **f)** 8 cm
3. **a)** Answers will vary. For example, 1 cm represents 10 cm. **b)** Answers will vary. For example, 1 cm represents 1 m. **c)** Answers will vary. **d)** Answers will vary. For example, comparing the size of the paper to the size of the original

8.7 Putting It All Together: Designing a Playground p. 155

Answers will vary.

8.8 Chapter Review pp. 156–57

1. 10 m
2. 10 feet
3. **a)** 100 square feet **b)** 9 **c)** 40 feet **d)** 4
4. outside: 53.4 mm; inside: 44 mm
5. **a)** 33 m^2 **b)** rectangular rug: 8 m^2; circular rug: 3.1 m^2 **c)** 21.9 m^2 **d)** 26 m
6. Answers will vary. For example, pace off, using about 1 m steps, rectangular portions, calculate the area of each, and add them together.
7. nine times
8. twice as long and twice as wide
10. **b)** 471.7 cm

Chapter 9: Measurement and 3-D Design

9.1 Rectangular Prisms pp. 160–63

Practise

1. Estimates will vary. **a)** 0.756 m^3 **b)** 2016 cm^3 **c)** 10 648 cubic inches
2. Estimates will vary. **a)** 5.1 m^2 **b)** 1152 cm^2 **c)** 2904 cm^2
3. **a)** $1944 **b)** $1504.32 **c)** $399.50
4. Answers will vary. For example, estimate its length, height, and width. The volume might be of interest to someone responsible for heating and cooling the building to determine the type of furnace and air conditioning to get.
5. **a)** 1.5 cubic yards **b)** $108
6. Answers will vary.
7. **a)** 265 square inches **b)** 337.5 square inches **c)** 72.5 square inches; the area of the wrapping paper minus the surface area of the stacked paper **d)** 27%
8. Answers will vary.
9. **a)** 4110 cm^3 **b) i)** 1600.1 cm^2 **ii)** 1613.8 cm^2 **iii)** 1846.7 cm^2 **iv)** 1998.8 cm^2 **c)** i) would have the least surface area. **i)** 600 cm^2 **ii)** 700 cm^2

9.2 Cylinders pp. 164–67

Develop

1. **a)** Yes, a cylinder is similar in shape to a rectangular prism, except the base is a circle. **b)** a circle, $A = \pi r^2$ **c)** $V = \pi r^2 h$
2. **a)** 3 **b)** circle, $A = \pi r^2$ **c)** rectangle, $A = lw$ **d)** ii **e)** surface area = $2(\pi r^2) + \pi dh$ or $2(\pi r^2) + 2\pi rh$
3. **a)** Answers will vary. **b)** Answers will vary. **c)** Answers will vary. **d)** The volume should be fairly close to the given capacity.

Practise

4. Estimates will vary. **a)** 624.2 cm^3 **b)** 1528.6 cm^3 **c)** 1524 cm^3

5. Estimates will vary. **a)** 407.2 cm^2 **b)** 751.3 cm^2 **c)** 783.4 cm^2

6. **a)** 0.88 m^3 **b)** 7 **c)** \$60.86

7. yes

8. 346 mL

9. **a)** the three-wick candle **b)** \$24. Answers will vary. For example, amount of labour required

10. **a)** **i)** yes, $V = 100.3$ cm^3 **ii)** yes, $V = 100.3$ cm^3 **iii)** yes, $V = 100.1$ cm^3 **b)** the bottle with a radius 2.5 cm and height of 5.1 cm. It is made of the least glass as its surface area is the least of the three.

11. **b)** volume of long tube = 63.2 cubic inches; volume of short tube = 81.8 cubic inches **c)** Volume is the product of π and three dimensions. Since the radius is used twice in the formula, $V = \pi r^2 h$, it has twice as much impact on the final result as the height does. So, when the radius is the longer side of a sheet of paper as in a short wide tube, it creates the greater volume.

12. 16. The volume of a round bale is 135 648 cubic inches, and the volume of a small square bale is 8512 cubic inches.

9.3 3-D Drawings p. 168

Develop

1. Answers will vary.

Practise

2–4. Answers will vary.

9.4 Scale Models pp. 169–70

Develop

1. **a)** 5 feet **b)** $7\frac{1}{2}$ feet

2. **a)** Answers will vary. For example, 1:20 **b)** Answers will vary. For example, 1:36 **c)** Answers will vary. For example, 1:300

Practise

3. 10.67 m

4. **a)** Centre Block width: 1.45 m; Centre Block height: 0.75 m; Peace Tower height: 0.55 m; ground to centre of clock height: 0.65 m; clock face diameter: 0.05 m; minute hand length: 0.025 m; hour hand length: 0.015 m; flag: 0.045 m by 0.02 m **b)** Yes, it is too large to carry.

5. **a)** ii **b)** 0.25 m long, 0.03 m wide, 0.07 m high, 0.02 wheel diameter **e)** actual volume: 65.625 m^3; model volume: 0.000 525 m^3. The actual engine's volume is 125 000 times that of the model. **f)** actual surface area: 135.5 m^2; model surface area: 0.0542 cm^2. The actual engine's surface area is 2500 times that of the model.

9.5 Career Focus: Hobby Store Clerk p. 171

1. Scale is the ratio of the actual dimensions of an object compared to those of its model.

2. The model uses measurements that are 24 times as small as the actual object.

3. so that the diorama is realistic

4. **a)** The scales reduce the model to a suitable size for sale. **b)** They do not convert easily to imperial measurements like the other scales. **c)** scales based on multiples of 10 because the metric system is based on multiples of 10

9.6 Putting It All Together: Finishing a Basement p. 172

Answers will vary.

9.7 Chapter Review p. 173

1. **a)** the amount of space an object occupies **b)** the sum of all the areas of all the surfaces of an object **c)** the ratio describing the relationship between the dimensions of an object and its model

2. box, 36.5 cm^3

3. box, 697.6 cm^2

5. **a)** a reduction **b)** 48 inches, or 4 feet **c)** $1\frac{3}{4}$ inches

6. amount of space: 48 134.7 m^3; area occupied: 4000 m^2; cost of bricks: \$16 348.80; cost of glass: \$920.85; scale: Answers will vary. For example, 1:100; building: 0.5 m × 0.8 m × 0.12 m; greenhouse diameter: 0.07 m; greenhouse height: 0.035 m

Chapter 10: Transformations and Design

10.1 Geometric Aspects of Design pp. 176–80

Practise

1. **Note:** Every shape that has turn symmetry has at least two lines of symmetry and thus shows at least two reflections. When turn symmetry exists, such as in the red circle with the diagonal bar, or "no" symbol, the line symmetries have not been mentioned.
a) The black stop sign has turn symmetry about a point at its centre. So, it shows a rotation about that point. The arrows have turn symmetry about a point centred between them. So, they show a rotation about that point. The red "no" symbol has turn symmetry about a point at the centre of its diagonal bar. So, it shows a rotation about that point. The sign itself has turn symmetry about a point at its centre. So, it shows a rotation about that point.
b) The cars have line symmetry about a vertical line between them. So, they show a reflection in that line. Each individual car has line symmetry about a vertical line through its centre. So, each shows a reflection in that line. The red "no" symbol has turn symmetry about a point at the centre of its diagonal bar. So, it shows a rotation about that point. The sign itself has turn symmetry about a point at its centre. So, it shows a rotation about that point.
c) The "triangles" show a dilatation about a point at their centre, the red "triangle" being an enlargement of the white one. The sign itself has turn symmetry about a point at its centre. So, it shows a rotation about that point.
d) The arrows have turn symmetry about a point centred between them. So, they show a rotation about that point. The other symbol has line symmetry about a vertical line through its centre. So, it shows a reflection in that line. The sign itself has turn symmetry about a point at its centre. So, it shows a rotation about that point.
e) The entire sign, including the symbols on it, has turn symmetry about a point at its centre. So, it shows a rotation about that point. Each arrow has line symmetry about a line through its centre. So, each shows a reflection in that line. Each vertical bar has turn symmetry about a point at its centre. So, each shows a rotation about that point. The three vertical bars together show a translation.
f) The entire sign, including the symbols on it, has turn symmetry about a point at its centre. So, it shows a rotation about that point.
g) Each portion of the road symbol has line symmetry about a line down its centre. So, each shows a reflection in that line. Each vertical bar in the road has turn symmetry about a point at its centre. So, each shows a rotation about that point. The two sets of three vertical bars in the road show a translation. The three vertical bars in each portion of the road together show a translation. The three waves together show a translation. The sign itself has turn symmetry about a point at its centre. So, it shows a rotation about that point.
h) The sign itself and all the individual squares have turn symmetry about points at their centres. So, they show rotations about those points. The sign, including the symbols on it, has line symmetry about a horizontal line through its centre. So, it shows a reflection in that line. The arrow has line symmetry about a horizontal line through its centre. So, it shows a reflection in that line.
i) The sign, including the symbols on it, has line symmetry about a vertical line through its centre. So, it shows a reflection in that line. The arrow has line symmetry about a vertical line through its centre. So, it shows a reflection in that line. The bump has line symmetry about a vertical line through its centre. So, it shows a reflection in that line. The sign itself has turn symmetry about a point at its centre. So, it shows a rotation about that point.
2. **a)** The "waterfalls" show a dilatation.
b) The entire logo, excluding the printing, has line symmetry about a vertical line. So, it shows a reflection in that line. Each stylized H or person has line symmetry about a vertical line though its centre. So, it shows a reflection in that line. The two stylized Hs or persons together have line symmetry about a vertical line between them. So, they show a reflection in that line. The cross and its surrounding design have line symmetry about a vertical line through the centre of the cross. So, they show a reflection in that line.
c) The two Ns show a translation. Each N has turn symmetry about a point at its centre. So, each

shows a rotation about that point. The half sun has line symmetry about a vertical line through its centre. So, it shows a reflection in that line.
d) The entire logo, excluding the printing, has line symmetry about a vertical line. So, it shows a reflection in that line. Each of the cross, the person, the half sun, the water, and the open book has line symmetry about a vertical line through its centre. So, each shows a reflection in that line.

3. a) two or more shapes identical in size, shape, and orientation **b)** two shapes identical in size and shape, but opposite in orientation, or a shape with a line of symmetry **c)** two or more shapes identical in size and shape, but not the same or the opposite orientation, or a shape with turn symmetry **d)** two or more shapes with the same shape and orientation, but not the same size

4. a) Each section of the border, that is, a portion with the horizontal white bars and the bent white bars with a diamond in the middle, translates onto the next section. The entire border has a horizontal line of symmetry about a line through its centre. So, it shows a reflection in that line. Each of the red diamonds, white diamonds, red circles, blue ovals, and white ovals, including what is inside them, has turn symmetry about a point at its centre. So, each shows a rotation about that point. The red and white diamonds together show a dilatation about a point at their centre, the red diamond being a reduction of the white one.
b) Answers will vary.

5. a) Answers will vary.
b) Answers will vary.

6. Answers will vary.

10.2 Investigating Design Using Technology pp. 181–84

Develop

4. The initial moves 2 units right and 3 units up.

5. a) moves 3 units up **b)** moves 3 units right **c)** moves 3 units down **d)** moves 3 units left **e)** moves 3 units right and 3 units up **f)** moves 3 units left and 3 units down **g)** moves 3 units left and 3 units up **h)** moves 3 units right and 3 units down

6. c) The image initial is the same size and shape as the original initial, but has the opposite orientation to it. The initials have line symmetry in the reflection line.

7. In both, the image initials have the same size and shape as the original initials, but the opposite orientation to them. When the line is through the original initial, the image and the original overlap, but when the line is outside, they don't.

8. a) iii **b)** iv **c)** i **d)** v **e)** ii

9. c) The image initial is the same size and shape as the original initial, but oriented differently.
d) Each image initial is turned 45° relative to its original. When all the rotations are made, the result is a shape with turn symmetry.

10. b) In both, the image initials have the same size and shape as the original initials, but different orientations to them. When the point is on the original initial, the image and the original overlap, but when the point is outside, they don't. When all the rotations are made for both, both result in shapes with turn symmetry.

11. c) The image initial is oriented the same way as the original initial, but is three times as tall and three times as wide.

12. c) The image initial is oriented the same way as the original initial, but is half as tall and half as wide. Both dilatations keep the image proportional to and oriented the same way as the original; however, the first enlarged and the second reduced.

Practise

13. a) translation, either $(-4, 5)$ or $(4, -5)$, depending upon which is the original triangle **b)** dilatation, either of two times about the origin or of 0.5 times about the origin, depending upon which is the original triangle **c)** rotation, either of 45° or $-45°$, depending upon which is the original triangle
d) reflection in the horizontal axis

10.3 Designing a Logo p. 185

Develop

1. Answers will vary.

Practise

2. Answers will vary.

10.4 Career Focus: Sign Painter p. 186

1. accuracy, precision with hands, orientation to detail, and ability to work with variety of materials
2. Answers will vary. For example, different people have different responsibilities, but they all are part of the same project, and working as a group can yield better results.
3. **a)** recycling: the rotation of bent arrows suggests continuity or reuse. **b)** railway crossing: rotation creates an X or cross. **c)** radiation: rotation creates a turning fan effect.
4. Answers will vary.

10.5 Tiling a Plane p. 187

Develop

1. **a)** yes **b)** yes **c)** yes
2. **a)** yes **b)** yes **c)** yes **d)** yes **e)** yes **f)** yes
3. all work together

Practise

4. **a)** yes **b)** yes **c)** yes
5. a and b
6. The sum of the angle measures of the vertices that come together is 360°.

10.6 Designing Involving Tiling Patterns p. 189

1–4. Answers will vary.

10.7 Putting It All Together: Designing Geometrically p. 190

Answers will vary.

10.8 Chapter Review p. 191

1. **a)** The entire tile and its design has turn symmetry about a point at its centre. So, it shows a rotation about that point. Each individual "flower" has line symmetry about a line through its centre lengthwise. So, each shows a reflection in that line. **b)** The circle design has turn symmetry about a point at its centre. So, it shows a rotation about that point. Each bar has turn symmetry about a point at its centre. So, each shows a rotation about that point. Each set of bars has turn symmetry about a point at its centre. So, each shows a rotation about that point. The set of three bars on the top left shows a translation. In the other sets of bars, the short bars alone show translations. The two long bars on the bottom left show a translation. **c)** The flooring pattern has line and rotational symmetry. The octagons can be translated, reflected, and rotated onto each other, as can the small squares. **d)** The logo has line symmetry about a vertical line though its centre. So, it shows a reflection in that line.
4. **a)** Answers will vary. For example, square, equilateral triangle, and regular octagon
   b) The vertices must be able to come together to complete 360° exactly.

Index

A

affordable housing, 100
apartment costs, 76
area, 144
assignment of a lease, 70
availability of apartments, 64

B

bar graph, 3
bias, 31
blended payments, 86
budget, 105

C

calculating area, 144
calculating perimeter, 144
calculating surface area of a cylinder, 165
calculating surface area of a rectangular prism, 161
calculating volume of a cylinder, 165
calculating volume of a rectangular prism, 161
Canada Mortgage and Housing Corporation, 88
Career Focus
  building superintendent, 75
  decorating store clerk, 136
  factory worker, 14
  flooring installer, 150
  furniture refinisher, 118
  hobby store clerk, 171
  medical office receptionist, 50
  real estate agent, 94
  sign painter, 186
  telemarketer, 39
changing one item in a budget, 115
circle graph, 7
CMHC, 88
column graph, 3
components of a household budget, 105
condominium, 84
constructing graphs, 10
costs of maintaining a home, 91
costs of maintaining an apartment, 77
creating designs involving tiling patterns, 189
cylinder, 164

D

designing a logo, 185
designs involving tiling patterns, 189
dilatation, 178
diorama, 171
double line graph, 6

E

empirical probability, 46
enlargement, 151
equally likely outcomes, 54
estimating area, 144
estimating capacities, 132
estimating distances, 130
estimating large numbers illustrated visually, 134
estimating perimeter, 144
experimental probability, 54

F

favourable outcome, 52
fixed expenses, 105

G

geometric aspects of design, 176
geometric transformation, 178

H

home costs, 91
household budget, 105

I

imperial units of linear measure, 127
interpreting data, 35
interpreting graphs, 2

L

land transfer tax, 89
landlord rights and responsibilities, 71
lease, 68
line graph, 6
line symmetry, 177
logo, 176
looking for a home, 84

M

making decisions based on probability, 48
measuring lengths, 127
metric system, 124
misleading graphs, 20
monthly apartment costs, 76
monthly budget, 108
monthly mortgage payment factor, 101
mortgage, 86

N

new home, 84

O

Ontario Rental Housing Tribunal, 67
organizing data, 35
outcome, 52

P

perimeter, 144
pie graph, 7
population, 30
possible outcome, 52
probability, 46
probability experiments, 54
property taxes, 92
Pythagorean theorem, 142

R

rectangular prism, 160
reflection, 178
reflectional symmetry, 177
renting an apartment, 67
representative sample, 30
resale home, 84
rights and responsibilities of landlords, 71
rights and responsibilities of tenants, 71
rotation, 178
rotational symmetry, 177

S

sample, 30
sampling techniques, 30
scale drawing, 153
scale model, 169
simulation, 56
Statistics Canada, 2
statistics in the media, 33
sublet, 70
surface area, 160
survey, 31
symmetry, 177

T

tenant rights and responsibilities, 71
tenant, 65
theoretical probability, 48
3-D drawing, 168
tiling a plane, 187
transformations using technology, 181
translation, 178
turn symmetry, 177

V

variable expenses, 105
volume, 160

Photo and Figure Credits

Chapter 1

(page 2) Jonathan Hayward/CP Photo Archive

(page 4, bottom) WWW.JOBFUTURES.CA: National Outlook to 2004, Where They Work, Type of Employment, Distribution by Age, Work Prospects, Earnings, Employment by History, Human Resources Development Canada. Reproduced with the permission of the Minister of Public Works and the Government Services Canada, 2002.

(page 5, top) WWW.JOBFUTURES.CA: National Outlook to 2004, Where They Work, Type of Employment, Distribution by Age, Work Prospects, Earnings, Employment by History, Human Resources Development Canada. Reproduced with the permission of the Minister of Public Works and the Government Services Canada, 2002.

(page 6) Reprinted with permission – The Toronto Star Syndicate.

(page 13) Neal Preston/CORBIS

(page 16) Ken Fisher/Getty Images

Chapter 2

(page 35, bottom) John Ulan/CP Photo Archive

Chapter 3

(page 53, bottom) Frank Gunn/CP Photo Archive

(page 58) DILBERT © UFS. Reprinted by permission.

Chapter 7

(page 134, bottom) Harvey Schwartz/MaXx Images, Inc.

(page 135, top) Stephan Poulin/Superstock Images

Chapter 8

(page 143) Helen Norman/CORBIS

Chapter 9

(page 169) Steve White/CP Photo Archive

(page 170) Tom Kitchen/firstlight.ca

Chapter 10

(page 176)
Canadian Red Cross logo reprinted by permission of the Canadian Red Cross Society

Bluewater District School Board logo reprinted by permission of the Bluewater District School Board

City of London logo reprinted by permission of The Corporation of the City of London

City of Niagara Falls logo reprinted by permission of the City of Niagara Falls

Olympic logo © IOC/International Olympic Committee

(page 179)
Halton District School Board logo reprinted by permission of the Halton District School Board

Waterloo District School Board logo reprinted by permission of the Waterloo Region District School Board

(page 180)
District School Board of Niagara logo reprinted by permission of the District School Board of Niagara

Halton Catholic District School Board logo reprinted by permission of the Halton Catholic District School Board

Near North District School Board logo reprinted by permission of the Near North District School Board

Algonquin & Lakeshore Catholic District School Board logo reprinted by permission of the Algonquin & Lakeshore Catholic District School Board

Registered Trademarks

Microsoft Excel®

Corel Quattro Pro®

MLS®